George Johnson

Lectures on Bright's Disease

With Especial Reference to Pathology, Diagnosis, and Treatment

George Johnson

Lectures on Bright's Disease
With Especial Reference to Pathology, Diagnosis, and Treatment

ISBN/EAN: 9783337155865

Printed in Europe, USA, Canada, Australia, Japan

Cover: Foto ©berggeist007 / pixelio.de

More available books at **www.hansebooks.com**

LECTURES

ON

BRIGHT'S DISEASE:

WITH ESPECIAL REFERENCE TO

PATHOLOGY, DIAGNOSIS, AND TREATMENT.

BY

GEORGE JOHNSON, M.D., F.R.S.

FELLOW OF THE ROYAL COLLEGE OF PHYSICIANS ; HONORARY FELLOW OF
KING'S COLLEGE, LONDON ; PROFESSOR OF MEDICINE IN KING'S
COLLEGE AND PHYSICIAN TO KING'S COLLEGE HOSPITAL.

LONDON :

SMITH, ELDER, & CO., 15 WATERLOO PLACE.

1873.

TO

HIS PAST AND PRESENT PUPILS

These Lectures

IN SUBSTANCE THE SAME AS HAVE OFTEN BEEN

ADDRESSED TO THEM WITHIN THE WALLS OF KING'S COLLEGE

Are Dedicated

BY THEIR SINCERE FRIEND AND WELL-WISHER

THE AUTHOR.

PREFACE.

In reprinting these Lectures from the 'British Medical Journal,' the author, profiting by the kind suggestions of some friends who have read the Lectures as they appeared in the journal, has made here and there some changes and additions. His object has been to give a clear and concise account of Bright's Disease, and, in doing so, to avoid all unnecessary details, and, as much as possible, all doubtful and disputed points of pathology. Theoretical questions have been referred to only so far as they appear to throw light upon the etiology, the pathology, the diagnosis, and the treatment of the various forms and stages of Bright's Disease. The doubtful and disputed questions relating to the minute anatomy and pathology of renal diseases have but little interest for the practitioner, and in these Lectures, which were especially addressed to students, controversial topics will be found to occupy only a very small space.

Savile Row:
  September, 1873.

CONTENTS.

a

LECTURE IV.

CHRONIC BRIGHT'S DISEASE WITH A LARGE WHITE KIDNEY.

LECTURE V.

CHRONIC BRIGHT'S DISEASE, WITH A LARDACEOUS OR WAXY KIDNEY.

LECTURE VI.

ALBUMINURIA NOT ASSOCIATED WITH WHAT IS COMMONLY UNDERSTOOD AS BRIGHT'S DISEASE.

LECTURE VII.

LECTURES

ON

BRIGHT'S DISEASE.

———◆◇◆———

LECTURE I.

The Minute Anatomy and Physiology of the Kidney—Physical Characters
of the Urine—The Mechanism of Albuminuria—Mode of Testing for
Albumen—Bright's Disease—History and Definition of the Term—
General Propositions relating to Bright's Disease: 1. It is of Constitu-
tional Origin: Proof of this—2. The Primary and Chief Changes occur
in the Gland-Cells of the Kidney—3. Changes in Basement-Membrane of
Tubes and Malpighian Capsules often misinterpreted—4. Changes in
Vessels of Kidney and other Organs later and less constant—5. Morbid
Products appearing as Tube-Casts in the Urine are of great diagnostic
value—Mode of examining Urinary Sediments.

AN EXACT KNOWLEDGE of the structure and functions of
the Kidney is essential for a correct interpretation of its
diseases. I therefore beg first to direct your attention to
certain points of anatomy and physiology which will be
found hereafter to have a direct bearing upon important
pathological questions.

A longitudinal section of the kidney shows it to be composed of a cortical and a medullary portion. The medullary
portion is arranged in the form of cones or pyramids—
pyramids of Malpighi—usually from twelve to fifteen in
number, the bases of which are directed outwards towards
the surface of the gland, becoming gradually continuous

B

with the cortical portion ; while the apices are directed
inwards towards the cavity or pelvis of the kidney. The
cortical portion occupies the entire surface of the organ,
forming a layer about two lines in thickness opposite the
bases of the medullary cones, and sending prolongations
inwards between the cones ; so that each medullary cone
is surrounded, except at its apex, by the cortical portion
of the gland. The kidney is a tubular gland. The tubes
of the cones take, for the most part, a straight course ;
while those of the cortex are extremely convoluted and
tortuous. Tracing the tubes from the apex of a medullary
cone, on the surface of which their open mouths may be
seen, they are found to take a straight course through the
pyramid, branching dichotomously, and diverging from
each other as they proceed. After reaching the base of
the pyramid, their course through the cortical portion
varies : many tubes immediately become very tortuous,
some of them bending down into the interpyramidal portions of the cortical substance, while others pass on in
sets and in straight lines towards the surface ; the tubes
on the sides of each bundle diverging successively, and
then taking a tortuous course through the cortical substance, so that only a few of the central tubes in each
bundle retain their straight course quite up to the surface
of the kidney. These all finally turn backwards, making
many convolutions in the cortical portion of the gland.
After leaving the medullary cones, the branching of the
tubes, except in very rare instances, appears to cease. In
all the numerous sections of the kidney that I have examined, I have never seen a convoluted uriniferous tube
either branching or anastomosing with another tube.
Some of the convoluted tubes dip down amongst the
straight tubes, forming loops with their convexities towards
the apex of the pyramid. Henle erroneously supposed that
these looped tubes were closed at both ends, and therefore
quite distinct from those which open into the pelvis of the

kidney. There is good anatomical evidence that, as each convoluted uriniferous tube at one extremity forms a globular dilatation, which constitutes the capsule of the Malpighian body, so at the other end it passes into a straight tube which opens into the pelvis of the kidney. For an excellent criticism of the wild and baseless speculations to which Henle's statements have given rise, I refer you to Dr. Beale's book on ' Kidney-Diseases,' etc. These speculations have no practical bearing upon the diagnosis or the pathology of renal diseases, and I shall not refer to them further.

We have next to trace the very remarkable arrangement of the blood-vessels within the kidney. The renal artery, entering the hilum of the kidney, sends small branches to the areolar and adipose tissue outside the pelvis, and then, passing into the substance of the organ, breaks up into terminal branches, which, with a few exceptions to be presently mentioned, correspond in number with the Malpighian bodies. Each terminal artery—the *afferent artery* of the Malpighian body—perforates a Malpighian capsule, and thus passes within the dilated end of the uriniferous tube. It there breaks up into loops of capillaries, which take a more or less tortuous course ; these again unite into a single *efferent vein*, which pierces the Malpighian capsule near the entrance of the afferent artery ; it then enters the capillary plexus which lies outside the tubes—the *intertubular plexus* (see Fig. 1). The intertubular capillaries anastomose freely on all sides, so as to form a continuous network, whence the blood is ultimately collected into the commencing branches of the renal vein. The course of the circulation through the kidney, then, is as follows:—The blood passes from the renal artery through the afferent arteries into the Malpighian capillaries ; from these it is carried by the efferent veins to the intertubular capillaries ; and thence it passes out of the kidney by the renal vein. The greater part of the blood which passes into the interior of the kidney takes the course which I have described; but amongst the straight tubes of the pyramids there are certain *vasa recta* which have a different distribution. Some of these *vasa recta* are efferent veins from Malpighian bodies near the bases of the pyramids, which, as originally described by Bowman, take a straight course towards the apices of the cones, and terminate in capillaries, from which the blood

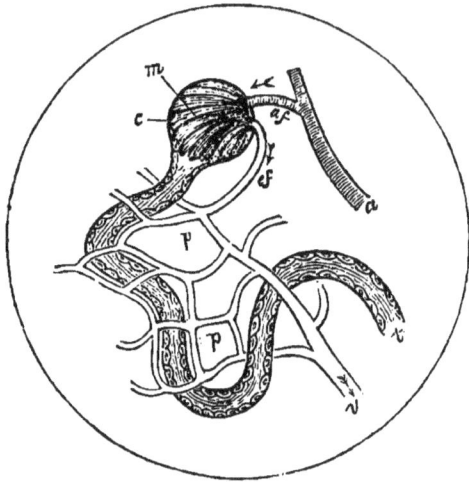

Fig. 1.—Plan of the Minute Structure of the Kidney. *a*. Artery, sending an afferent branch, *a f*, which breaks up into *m*, the Malpighian capillaries. *e f*. Efferent vessel, which conveys the blood from the Malpighian capillaries into *p*, the plexus of capillaries between the tubes. These again unite, and form the vein *v*. *t*. The uriniferous tube, *c*. The capsule of the Malpighian body. The course of the circulation is indicated by the arrows.— × 60.

is returned by venous radicles, which also take a straight course and join the renal vein. But, in addition to these venous branches, it has been shown by Virchow, Beale, and others, that there are arterial *vasa recta* which pass off from the artery, take a straight course between the tubes of the cones, and terminate in a capillary network surrounding the tubes. These arterial *vasa recta* are probably the chief nutrient vessels of the pyramids. They may, therefore, be looked upon as analogous to the bronchial arteries in the lungs and the hepatic artery in the liver.

Each Malpighian body, as we have seen, consists of a globular plexus of capillaries contained within the dilated end of a convoluted tube; and we have now to consider briefly the structure of the uriniferous tubes. The tube is composed of two anatomical elements—the *basement-membrane* and the *epithelium.*

The basement-membrane is a thin transparent lamina, appearing, as a rule, structureless and quite homogeneous, the slightly fibrous appearance which it sometimes presents being probably due to contraction and corrugation after the sections which must of necessity be made for microscopic examination. This membrane is in direct contact on its outer surface with the intertubular capillaries, and on its inner surface with the epithelial lining of the tubes.

The epithelium in the convoluted tubes differs from that in the straight tubes. In the convoluted tubes, the epithelium is of the true glandular character. The cells are somewhat angular in outline; and between this and the central nucleus there are a number of granular particles. The cell-wall is often indistinct, and readily disintegrated by the action of water. The cells form a single layer within the tubes; and this cell-lining occupies from one-third to one-half of the diameter of the tube, leaving a clear canal in the central axis of the tube. (Fig. 2.)

The epithelium in the straight tubes is flatter, less granular, and has more the character of pavement-epithelium; so that the clear canal within a small straight tube is wider than that of a convoluted tube of larger size.

In the human kidney no epithelial cells can be seen either lining the Malpighian capsule or covering the Malpighian capillaries. In the kidney of the rabbit and other of the lower animals, after staining with nitrate of silver and carmine, an endothelial lining of the capsule is rendered visible. In the kidney of the newt and the frog, a delicate layer of ciliated epithelium may be seen within that portion of the Malpighian capsule which lies next to the opening of the tube; and, in the newt's kidney, vibra-

Fig. 2.—Portion of a Convoluted Uriniferous Tube. The lining of glandular epithelium leaves a clear canal in the middle, which is equal to about half the diameter of the tube.— × 200.

tile cilia may be seen throughout the entire length of the uriniferous tubes.

The appearance to which Goodsir originally gave the name of the *matrix* of the kidney has been a source of much perplexity to anatomists and pathologists. Fig. 3.

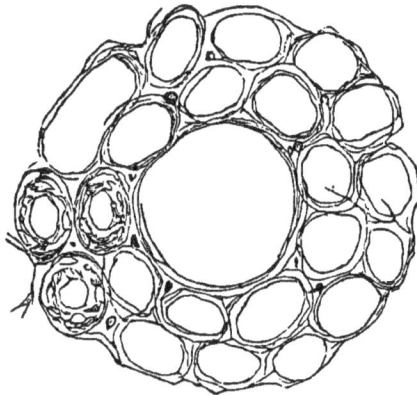

Fig. 3.—Section of the Cortex of the Kidney, after washing in water to remove the Gland-cells. The smaller rings are sections of the basement-membrane of the tubes; the larger ring is a section of a Malpighian capsule. In three sections of a tube the gland-cells remain. Sections of capillaries are seen here and there in the angular spaces between the tubes.— × 200.

represents an appearance which results from washing a.

thin section of the cortex of an uninjected kidney in water,
so as to remove the gland-cells. The appearance is that
of a fibrous network enclosing circular and oval spaces.
The explanation of the appearance is this. The tubes lie
in close contact with each other, having the intertubular
capillaries between them. A thin transverse section gives
a reticular appearance; the rings being formed by the
basement-membrane of the tubes, with the capillaries in
the interspaces and angles. The so-called matrix has no
existence apart from the basement-membrane and capillaries. The convolutions of the tubes and the network of
capillaries mutually support each other. No connective
or supporting tissue is required; and, as Dr. Beale well
remarks, the intervention of any such tissue would tend to
increase the distance between the secreting cells and the
blood, and so render the gland less perfectly fitted for the
discharge of its function. Ludwig states that 'no fibrillated connective tissue exists between the tortuous portions
of the urinary tubules' (Stricker's 'Manual of Histology,'
New Sydenham Society's Translation, vol. ii. p. 106). There
is no more appearance of connective tissue on the *outer*
surface of the basement-membrane between it and the
capillaries than there is on the *inner* surface between it
and the gland-cells. The tissues on either surface of the
basement-membrane adhere to it without the intervention
of another tissue to which the term connective tissue can
be given (see Fig. 4). If there be any connecting medium
it is a homogeneous and structureless element.

You may make a coarse imitation of the fibrous network
of the kidney by taking half a dozen india-rubber tubes,
cementing them together side by side, so as to form a
bundle of parallel tubes. Transverse sections will then
form a network, the rings of the meshes being formed by
the divided india-rubber tubes, as the reticular appearance
in the kidneys is the result of sections of the basement-
membrane of the uriniferous tubes.

Bear in mind, then, that there is no distinct structure to which the term 'matrix' can be applied. The fibrous appearance represented in Fig. 3, which has been often

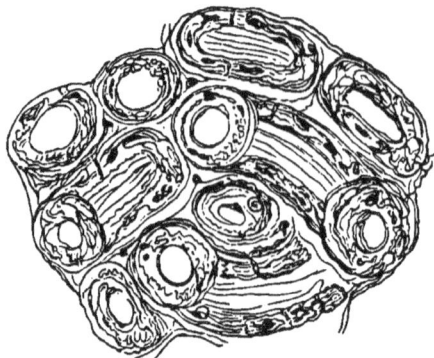

Fig. 4.—Section of the Cortex of the Kidney. The gland-cells are here attached to the inner surface of the basement-membrane. The light interspaces correspond with the rings of the basement-membrane in Fig. 3.—× 200.

described as a morbid formation of fibrous tissue surrounding and constricting the tubes, is, I hope, rendered quite intelligible by the description which I have given you.

The diameter of the convoluted tubes is remarkably uniform, and equals about one-five-hundredth of an inch.

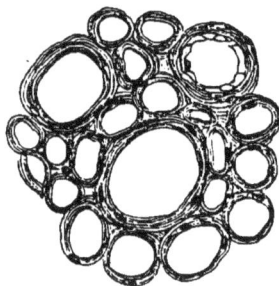

Fig. 5.—Section of a Medullary Cone. The rings which are here of unequal sizes are sections of the straight tubes. In one section the epithelium remains.— × 200.

That of the straight tubes is much more variable. While many straight tubes have a narrower outline than the convoluted tubes, but a wider canal, in consequence of the more flattened form of the epithelium, others, especially near the apex of a cone, are more than twice as large as the convoluted tubes. This may be seen by a comparison of Figs. 3 and 5, the latter representing a section of straight tubes in a cone. The basement-membrane of the straight tubes is somewhat thicker than that of the convoluted tubes, and the medullary cones are firmer than the tissue of the cortex.

Nerves.—The nerves of the kidney are chiefly derived from the sympathetic. In man and in the higher animals it is difficult to trace their distribution; but in the kidney of the newt Dr. Beale has found that not only are the terminal branches of the nerves distributed to the small arteries and veins, but also to the convoluted tubes and to the Malpighian and intertubular capillaries. The nerve-fibres are all connected with ganglion-cells, from each of which two or more fibres proceed in different directions, and so establish a communication between various parts of the organ. It is probable, as Dr. Beale suggests, that the nerves which are distributed over the uriniferous tubes constitute an afferent system, which, through the nerve-centres and the efferent nerves distributed to the arteries, are capable of influencing and regulating the blood-supply to the capillaries, and so the functional activity in health and in disease.

Practically, the kidney may be said to be made up of two sets of tubular vessels—one set of tubes containing blood, the other containing gland-cells ; and the organ is so constructed as to bring the two sets of tubes—the sanguiniferous and the uriniferous—into close and intimate relationship with each other.

The Function of the Kidneys is to discharge from the body superfluous water, together with certain peculiar urinary solids. There appears no reason to doubt the essential accuracy of Mr. Bowman's original doctrine that, while the convoluted tubes, with their lining of gland-cells,. are the agents by which the solids of the urine (the urea,. uric acid, etc.) are secreted, the watery portion of the secretion is chiefly discharged through the Malpighian bodies.

The convoluted tubes resemble, in all essential points, the secreting tissues of true glands, and especially in the character of their epithelial cells ; while the Malpighian bodies, in their structure and arrangement, form a striking contrast. The epithelial cells either cease altogether or entirely change their character within the Malpighian capsules. The Malpighian capillaries lie within the dilated ends of the tubes, and are entirely uncovered by epithelium. ' It would be difficult,' as Mr. Bowman says, ' to conceive a disposition of parts more calculated to favour the escape of water from the blood, than that of the Malpighian body.' Each afferent artery breaks up into a number of minute capillaries of far greater aggregate capacity than itself. Hence must arise an abrupt retardation of the blood-stream. The vessels in which this delay occurs are uncovered by cells. The interior of the capsule certainly in the lower animals, and probably in the higher, is lined by cilia whose motion directs the current of liquid towards the orifice of the tube. ' Why is so wonderful an apparatus placed at the extremity of each uriniferous tube, if not to furnish water, to aid in the separation and solution of the urinous products from the epithelium of the tube ? '

The epithelium of the straight tubes, as I have before mentioned, is allied to the lamelliform or pavement variety. It probably has no glandular function, the tubes which form the medullary cones being merely ducts for conveying away the secreted products from the convoluted tubes into the pelvis of the kidney.

The precise mode in which the glandular epithelium separates its peculiar products from the blood and discharges:

them into the duct, is a mystery which has not yet been solved. It is probable that the cells of the kidney continually pass away in the secretion, and that they are as constantly replaced by new formations ; but, whatever may be the process by which these changes are effected, no entire gland-cells, nor even the *débris* of renal epithelium, are normally visible in the urine. The appearance of renal cells in the urine affords undoubted evidence of a pathological process.

Conflicting results have been obtained by different experimenters in their attempts to solve the question whether the peculiar urinary constituents exist ready formed in the blood and are only separated by the kidney, or whether they are formed wholly or in part by the gland. It had long been the accepted doctrine that urea and uric acid exist normally in the blood, that they are thrown out by the kidneys, and that they accumulate and cause uræmia when the secretory function of the kidney has been impaired by disease. Dr. Oppler of Berlin threw a doubt upon this doctrine. He found, as he believed, much more urea in the blood of dogs whose ureters had been tied than in the blood of those whose kidneys had been extirpated, and he concluded that the excess was due to the formation of urea by the kidneys in the first class of cases. It is probable that animals live longer after ligature of the ureters than after the more formidable operation of nephrotomy, and this may explain the excess of urea in the ligatured cases. The more recent observations of Meissner and others have tended to re-establish the older doctrine by showing that urea and uric acid exist in the blood of healthy animals, and, moreover, that they are so abundant in the liver as to render it probable that the liver is the chief seat of their formation.[1]

[1] See for references to these experiments, New Sydenham Society's *Biennial Retrospect*, 1867–8.

There is, then, good reason for the doctrine that the urinary constituents are largely brought to the kidney by the blood, whence they are discharged through the uriniferous tubes of the gland; and hence arises the contamination of the blood by urine when the kidneys are structurally changed and their excretory function suspended or much impaired.

Physical Characters of the Urine.—Healthy urine is a transparent sherry-coloured liquid, having an acid reaction and a density usually ranging between 1015 and 1025, but it may temporarily fall much below or rise considerably above these limits without being morbid. The daily secretion of urine has been estimated by some observers to be as low as 35 ounces; by others as high as 81 ounces (Parkes 'On the Urine,' p. 5); the mean being $50\frac{1}{2}$ ounces. The amount secreted depends upon the measure of fluid taken in and the amount passed off by other channels, especially by the skin.

Referring you for a detailed account of the chemistry of the urine to the works of Parkes, Thudichum, Beale, etc., I may remind you in passing that, as the lungs and the liver are large eliminators of carbon, so the urinary secretion is remarkable for the abundance of its nitrogenous constituents. Urea, the chief urinary solid, contains a large proportion of nitrogen, and the amount of urea discharged by an adult male in twenty-four hours ranges, according to different observers, from 286 grains to 688 grains, the mean being 512 grains (Parkes, pp. 7 and 8).

The Mechanism of Albuminuria.—Now, before I proceed further, let me show you, by referring to the anatomy of the kidney, that the peculiar position of the Malpighian capillaries within the dilated ends of the uriniferous tubes is attended with this result, that any interference with the circulation through the kidney is apt to be associated with an escape of blood-constituents through the Malpighian capillaries, which, mingling with the urine, render it

either bloody, or, if the serum alone escape, simply albuminous.

Looking at the plan of the renal circulation (Fig. 1),
you see that, whether the escape of blood-constituents be
traceable to an altered physical relation between the blood
and the walls of the vessels, or to engorgement of the
Malpighian capillaries, the result of an increased afflux of
blood through the arteries or of an impeded return of
blood through the intertubular capillaries and veins consequent on an obstruction within the kidney itself, or
beyond, in the heart or lungs—in each and every case the
blood-materials, transuding through the walls of the Malpighian capillaries into the tubes, mingle with the urine
and render it bloody or albuminous. There are many
interesting points of analogy between the liver and the
kidney as regards structure, functions, and pathology;
but in the liver there is nothing analogous to the intratubular Malpighian capillaries, and therefore, while albuminuria is of very common occurrence, an albuminous or
sanguineous condition of the bile is a rare event.

Tests for Albumen in the Urine.—Albuminous urine is
usually coagulated by heat short of the boiling point, and
by nitric acid. A careful application of both tests can rarely
lead to error, but mistakes have often arisen from the
employment of only one of these methods. Heat alone will
not coagulate albumen in urine which is neutral or alkaline.
In such a case, the addition of nitric acid coagulates and
precipitates the albumen. Add a few drops of liquor
potassæ to albuminous urine, and you will find that it will
not coagulate by heat.

On the other hand, in urine which is alkaline, neutral,
or feebly acid, a precipitate of phosphatic salts may be
thrown down by boiling, and this may be mistaken for
albumen. The addition of a drop or two of nitric acid
immediately dissolves the phosphatic sediment and renders
the urine clear. You see, then, that if you trust to heat

alone, you may fail to detect albumen, which is abundantly present; or, on the other hand, you may mistake a phosphatic sediment for albumen.

The addition of a few drops of nitric acid to albuminous urine usually forms an unmistakable coagulum. This test employed alone would less frequently mislead than heat alone. In using the nitric acid test, certain facts are noteworthy, and some care is required to avoid error.

1. The addition of a drop or two of nitric acid occasionally forms a precipitate of albumen, which is rapidly redissolved; then the further addition of the acid causes a permanent precipitate.

2. The addition of a small quantity of nitric acid to urine—as much acid, for instance, as may be left in an unwashed test-tube which has contained a mixture of urine and nitric acid—will prevent the coagulation of albumen by heat.

3. Heat alone does not always coagulate albuminous urine which is highly acid, and to which no extraneous matter has been added.

It has been found that any free acid, whether vegetable or mineral, which does not itself coagulate albuminous urine will prevent the coagulation of albumen by heat.

4. An excess of nitric acid, may decompose or dissolve a scanty sediment of albumen, especially when added to boiling urine.

5. Nitric acid added to urine containing urates abundantly may cause a cloudy precipitate of uric acid. When boiled, the urine becomes clear and of a dark colour by the decomposition of the uric acid.

6. A crystalline precipitate of nitrate of urea in concentrated urine could scarcely be mistaken for albumen.

7. Urine containing copaiba, cubebs, and other resinous substances, may be rendered turbid by nitric acid, but not by heat. The odour of the resin may be detected in such urine.

In conclusion, let me remind you of two delicate tests for a mere trace of albumen : 1, in urine which is turbid with urates ; and 2, in clear urine.

1. Pour the turbid urine into a clear test-tube until the tube is two-thirds full. Hold the tube steadily by the lower end, and heat the upper stratum of urine over a spirit-lamp. The liquid is first cleared by solution of the urates, and then made opalescent by coagulation of the albumen.

2. Pour the clear cold urine into a test-tube until it is half full ; slope the tube, and allow from five to ten drops of nitric acid to trickle down the under side of the tube : three layers are quickly formed—clear nitric acid below, clear urine at the top, and an opalescent stratum of slightly coagulated urine between the two.

Definition of Bright's Disease.—Having made your-selves acquainted with the structure and functions of the kidney, you are prepared to enter upon the study of its diseased conditions ; and I now proceed to give you some account of a most important and interesting class of cases which are usually included under the name of Bright's Disease. The history of this term may be very briefly told. Before the time of Dr. Bright, it was known that dropsy and disease of the kidney were sometimes associated. It was also known that some dropsical patients had albu-minous urine. (See 'Observations on the Nature and Cure of Dropsies,' by John Blackall, M.D., 3rd ed., 1818.) Dr. Bright's great merit and originality consisted in this, that he pointed out the frequent association of dropsy and albuminuria with very striking pathological changes in the kidney. In the first volume of his 'Reports of Medical Cases,' published in 1827, he described and represented by beautiful coloured drawings various morbid appearances in the kidney ; some kidneys being large and congested ; others large and anæmic ; and others, again, contracted and granular. He showed that these forms of renal disease

are of every-day occurrence ; that they are frequently associated not only with dropsy, but with many other formidable secondary diseases; and thus he opened up the
great field of renal pathology, which had previously been,
for all practical purposes, an almost unknown region.
These morbid conditions of the kidney having been made
known, it became necessary to give them a name, and
various names have been proposed. Rayer used the term
' nephrite albumineuse ' to designate this class of diseases.
The objections to this term are, first, that every form of inflammation of the kidney may be associated with albuminous
urine ; and second, that some forms of the disease under
consideration are not of an inflammatory nature. Dr. Christison called the disease ' granular degeneration.' The kidneys, it is true, are often granular ; but in some of the
most characteristic cases they are quite smooth, and not at
all granular. Each of these terms, then, being insufficient
and objectionable, it has become the custom to designate
the morbid states of the kidney by the name of the distinguished physician who discovered them ; and so the
term ' Bright's disease ' has come into very general use
both in this country and abroad. The term is sufficiently
convenient and unobjectionable, if only we can agree upon
a definition. The designation Bright's disease seems to
involve the idea of unity ; and some pathologists have
maintained that all the morbid changes in the kidney to
which attention was directed by Dr. Bright are the result
of a single morbid process in different stages and of various
grades of intensity. I shall have no difficulty in convincing you that this view is erroneous. Meanwhile, however, I must ask you to bear in mind that under the name
of Bright's disease are included various forms of acute and
chronic disease. In the new nomenclature of the Royal
College of Physicians, the term is thus explained : 'Bright's
Disease. *Synonym :* Albuminuria. *Definition :* A generic
term including several forms of acute and chronic disease

of the kidney, usually associated with albumen in the urine, and frequently with dropsy, and with various secondary diseases resulting from deterioration of the blood.'

Accepting this definition of Bright's disease, we shall find that it is nearly synonymous with albuminuria—nearly, but not quite. For, on the one hand, in some quite exceptional cases, both acute and chronic, albuminuria is sometimes absent; and, on the other hand, albuminuria may be unassociated with Bright's disease. For example, the mixture of blood or pus with the urine of course renders it albuminous; but hæmaturia and purulent urine, although often associated with Bright's disease, may result from other and quite distinct pathological conditions, either general or local. And, again, in the advanced stages of valvular disease of the heart, and in some cases of extreme emphysema of the lungs with bronchitis, albuminuria may be caused by passive congestion of the kidney resulting from an impeded circulation through the heart and lungs, and a consequent engorgement of the whole systemic venous system; yet albuminuria thus originating from purely mechanical causes would not be correctly designated a form of Bright's disease. With these limitations, however, the terms 'albuminuria' and 'Bright's disease' may be looked upon as practically synonymous; and, to avoid wearisome reiteration, I shall employ sometimes one and sometimes the other term.

Now let me impress upon you that, according to the Nomenclature and definition of the Royal College of Physicians, Bright's disease is not always and of necessity a hopelessly incurable malady. Under this common designation will be included on the one hand cases as curable as a simple catarrh or a slight pneumonia, and on the other hand cases as intractable as advanced pulmonary phthisis. The first great division of cases of Bright's disease is into acute and chronic; and, in any case that comes under your notice, there always arises this most important practical question, Is the disease acute, and therefore probably curable? or is the case one of chronic and advanced degeneration of the kidney, and therefore probably irremediable? A careful study of the entire history of the disease, and of each particular case that comes under your observation, will alone enable you to give a true and trustworthy answer to this question.

Before I proceed to discuss the various forms of Bright's disease, I wish to direct your attention to certain general propositions which are true of all forms of the disease.

Proposition I. Bright's disease is not a merely local malady, but a disease of constitutional origin; and the proximate cause of the renal disease is, in all probability, a morbid condition of the blood.

The proofs of the blood-origin of Bright's disease are to be found in the entire physiological history of the disease. Much of this evidence will come under our consideration hereafter; but some facts bearing upon the question may with advantage be referred to now.

First, then, the disease is a bilateral disease. The rule is, that both kidneys, receiving the same morbid blood, are both affected, and both by the same form of disease, although the degeneration is sometimes more advanced in one kidney than in the other. The exceptions help to prove the rule. For example, one kidney may be absent or undeveloped; or it may have been destroyed by an abscess or by the impaction of a calculus. Bright's disease occurring in such cases would of necessity be unilateral. But the most instructive case of unilateral Bright's disease that I am acquainted with has been published by Dr. Moxon, in the 'Pathological Transactions' (vol. xix. p. 268). In a woman aged 34, who died of dropsy, the right kidney had the characters of a large, pale, granular Bright's kidney. 'The left, on the contrary, was rather small, and of the colour and appearance of a healthy kidney.' A microscopic examination showed the large kidney to be much diseased; the smaller 'practically healthy.' The explanation of this remarkable difference was found in the fact that the left renal artery was plugged by a very old fibrinous coagulum, probably derived from the interior of the heart. Dr. Moxon, in his interesting comments on this case, suggests that, while one kidney was saved by a diminution or rather a suspension of its function, the other was destroyed by an excess of function. We know, as he says, that an excess of normal function (as, for instance, when one kidney is destroyed by an impacted calculus) causes not Bright's disease, but simple hypertrophy of the kidney. In this case, the result was to aggravate a disease which probably had already commenced. But, with reference to the theory of blood-poisoning, which I am now endeavouring to illustrate, I would suggest that the fibrinous plug saved the left kidney by excluding morbid blood, while it damaged the right by diverting to it a double supply of the same morbid blood. This case, therefore, confirms and helps to explain the rule that Bright's disease is bilateral. The bare nutrition of the left kidney in the case referred to was probably maintained by anastomoses between the renal artery and other branches from the aorta. The kidney may be partially injected from the aorta after ligature of the renal artery. 2. Confirmatory evidence of the blood-origin of Bright's disease is derived from the fact that the malady occurs in association with constitutional states in which a morbid condition of blood may confidently be assumed to exist. Albuminuria, varying in degree and in duration, has been found more or less frequently associated with scarlet fever, diphtheria, measles, small-pox, erysipelas, pyæmia, typhus and typhoid fever, rheumatic fever, malarious fevers, cholera, purpura, scurvy, diabetes, syphilis, certain forms of pneumonia, pregnancy, the absorption of secretions from the interior of the uterus after parturition, gout, the abuse of alcoholic liquors, excessive eating, certain forms of dyspepsia resulting, as may be

supposed, in the passage of crude materials into the circulation, a poor and insufficient diet, purulent and other
exhausting discharges, and, lastly, suppressed action of
the skin by exposure to cold, and especially to cold and
wet combined.

Proposition II. The morbid blood, which is assumed to
be the proximate cause of Bright's disease in all its forms,
exerts its influence primarily and chiefly upon the gland-
cells which line the convoluted tubes.   Look at a Bright's
kidney, or at one of Bright's beautiful plates, and you see
at a glance that the cortex or secreting portion of the
kidney is the seat of the disease, while the medullary cones,
even in the advanced stages of the malady, are left comparatively intact.   Analyse the diseased gland with the
microscope, and you find that the morbid process has
been concentrated, and, as it were, focused upon the
secreting cells within the uriniferous tubes.   The kidney,
as a great blood purifier, forms an outlet and a means of
escape for many useless and noxious materials which have
been developed within the system or introduced from
without; and in the discharge of this excretory function,
the gland undergoes the degenerative changes which are
designated Bright's disease.   The blood in the vessels of
the kidney probably contains no more noxious materials
than an equal volume of blood in any other tissue—that of
the voluntary muscles, for instance; but, during the process
of excretion, these products are withdrawn from the blood
and concentrated within the gland-cells of the kidney,
where they effect the morbid changes in question.

Proposition III. The structural changes which occur in
the basement-membrane of the tubes and in the Malpighian
capsules are direct results of the intratubular cell-changes.

These changes in the basement-membrane are often very
obvious and striking, but they have frequently been misinterpreted.   For example, thickening and corrugation of
the membranous walls of the tubes have been mistaken by

Virchow and his followers for a formation of connective tissue *between* the tubes, and thus all the phenomena of the disease have been misunderstood.

Proposition IV. During the progress of chronic Bright's disease, the blood-vessels in the kidney and in many other tissues and organs undergo very interesting changes, but these occur later and less constantly than those which affect the secreting tissues of the gland.

Proposition V. The pathological products of the structural changes within the tubes, being carried out by the liquid secretion, escape with the urine and appear in the form of cylindrical casts of the uriniferous tubes ; and a microscopical examination of these tube-casts affords most interesting and valuable information as to the nature and the stage of the renal disease. I shall hereafter show you the various forms of tube-casts, and explain to you their diagnostic significance.

Let me remind you that the examination of urinary sediments of all kinds is much facilitated by allowing the urine to stand for a few hours in a four-ounce conical glass. Then a portion of the sediment is to be taken up with a pipette and put into a glass cell, and this, covered with thin glass, is placed beneath a quarter-inch object-glass. The most convenient cell is made by cementing with marine glue a circular flat ring of glass upon the ordinary microscopic slip of glass. These cells are sold by all microscope makers. One cell, with ordinary care, will last for months. It is a waste of time to hunt for tube-casts in a drop of urine placed between two flat pieces of glass. The glass cell before mentioned holds several drops of sediment, and therefore greatly facilitates the investigation.

LECTURE II.

ON ACUTE BRIGHT'S DISEASE.

Synonyms—General Symptoms—Microscopic Appearances in the Urine—
Morbid Anatomy of the Kidney—Physiology of the Morbid Process—
Varieties of Acute Bright's Disease: 1. With Epithelial Desquamation
(Desquamative Nephritis)—2. Without Desquamation, and with or with--
out small Hyaline Casts—3. With Exudation-cell Casts, with or without
Epithelial Desquamation—4. Without Albuminuria—Changes in the
Blood—Etiology—Diagnosis—Prognosis.

I HAVE told you that the first great division of cases of
Bright's disease is into acute and chronic, and I now proceed to give you the pathological history of acute Bright's
disease. The synonyms for acute Bright's disease, given
in the Nomenclature of the Royal College of Physicians,
are 'acute albuminuria,' 'acute desquamative nephritis,'
and 'acute renal dropsy.' The renal disease associated
with dropsy, which often occurs in connexion with scarlet
fever, may be taken as a type of acute Bright's disease. I
will first give you a sketch of the ordinary course of this
form of disease, and I will then point out to you the chief
varieties and modifications of the malady.

The attack is usually ushered in by a sense of chilliness,
which may amount to actual rigors; a quick and throbbing
pulse, a hot and dry skin, a dry and coated tongue, thirst,
loss of appetite, pain in the back and limbs, headache,
and restlessness. In some cases, frequent vomiting occurs
at the commencement of the attack. In most instances,
dropsy is a very early symptom; the patient's attention,
or that of his friends, being arrested by an appearance of
unusual pallor and puffiness of the face, and especially of

the eyelids ; the swelling soon becomes general, affecting the subcutaneous areolar tissue throughout the body, and often one or more of the serous cavities. The urine is at this stage more or less scanty, occasionally almost or even altogether suppressed; usually it is dark coloured from admixture with. blood, the colour varying from a slight smokiness to a deep blood tinge, and it contains so large an amount of albumen as to become nearly solid with heat and nitric acid. The specific gravity varies considerably, being as often above as below the normal point. There is usually more or less pain and tenderness in the loins ; the pain is sometimes, though rarely, severe, and occasionally it extends to the inside of the thighs and to the testicles. There is frequent desire to pass water, and sometimes a sense of pain or scalding in the urethra. There is often more or less of uneasiness in the epigastrium, with flatulent distension of the stomach, especially after food, and nausea and vomiting are of common occurrence. In some cases, inflammation of one or more serous membranes—the pleura, the pericardium, or the peritoneum—occurs; or respiration is impeded by an œdematous or inflammatory effusion into the pulmonary air-cells and the smaller bronchi ; or the head-ache, which is often present from the commencement, becomes more severe, and is followed by one or more attacks of convulsions, from which the patient may recover, or which may be followed by fatal coma.

When the progress of the disease is favourable, one of the earliest signs of amendment is an increased secretion of urine. It may be that for some days only a few ounces of urine have been passed in the twenty-four hours ; it soon becomes more copious, of paler colour, of lower specific gravity and less albuminous, and the amount secreted may be as much as from four to six pints, the dropsy meanwhile diminishing daily. After an interval, varying from a few days to a month or more, the secretion of urine is reduced to the normal amount, the sediment diminishes, and at length disappears, the urine gradually resumes its normal colour and ceases to be albuminous. At any time during the convalescence, there may be a temporary increase of blood and albumen, with a diminished secretion of urine and a return of dropsy, if the congestion of the kidneys be increased by exposure to cold or by errors of diet. In most cases the dropsy disappears for days, and often for weeks or even months, before the urine has ceased to be albuminous. In some cases, although the dropsy and the pallor of the skin pass away, the urine remains albuminous, and the renal disease passes into a chronic form.

Acute albuminuria is sometimes unassociated with dropsy from its commencement to its termination. The terms 'acute albuminuria' and 'acute renal dropsy' are, therefore, not strictly synonymous.

Microscopic Appearances in the Urine.—A portion of

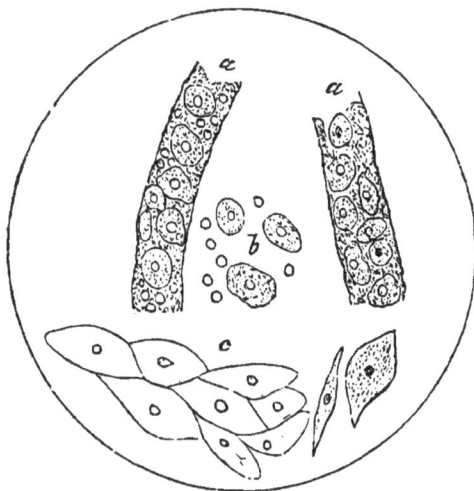

Fig. 6.—*a a.* Epithelial Casts. Casts of the Uriniferous Tubes entangling Renal Epithelium and Blood-corpuscles. *b.* Scattered Renal Gland-cells and Blood-corpuscles. *c.* Pavement-epithelium from the Vagina. This is broader, flatter, and less granular than the renal epithelium.— × 200.

the urinary sediment in the early stage of acute Bright's

disease being placed under a quarter-inch object-glass, presents very characteristic appearances. The most striking objects are solid cylindrical moulds of the uriniferous tubes. The basis of all renal tube-casts is fibrine which has coagulated within the uriniferous tubes, but these casts assume various appearances according to the nature of the products which they contain and the condition of the tubes in which they have been moulded. The casts which are most characteristic of acute Bright's disease are ' epithelial casts' (Fig. 6). These casts contain gland-cells evidently derived from the convoluted tubes of the cortex; they also entangle blood-corpuscules, and some casts are entirely composed of coagulated blood; these are called ' blood-casts' (Fig. 7). Together with the tube-casts, many scat-

Fig. 7.—Blood-casts composed of Fibrine entangling Blood-discs.—× 200.

tered renal gland-cells and blood-discs may usually be seen. In addition to the epithelial and blood-casts, we find in most cases of acute Bright's disease some small and large hyaline casts (Fig. 8). The difference between a small and a large hyaline cast is readily explained by referring to Fig. 2 in my previous lecture. The small casts, which are composed of pure fibrine, are moulded within the canal formed by the gland-cells, which retain their normal position within the uriniferous tubes; the large casts, on the other hand, are formed within tubes whose gland-cells have been removed. The diameter of these casts, therefore, is about twice that of the small casts, and equals that of the tubes whose basement-membrane constitutes the mould in which they have been formed. The larger casts may be simply hyaline, or they may entangle here and there a cell-nucleus or the fragment of a cell.

In cases of acute Bright's disease, the small hyaline casts

are often present in considerable numbers, while the large
hyaline casts are usually less numerous, and may be entirely
absent. On the other hand, there are unquestionably acute
and curable cases in which the large hyaline casts are very

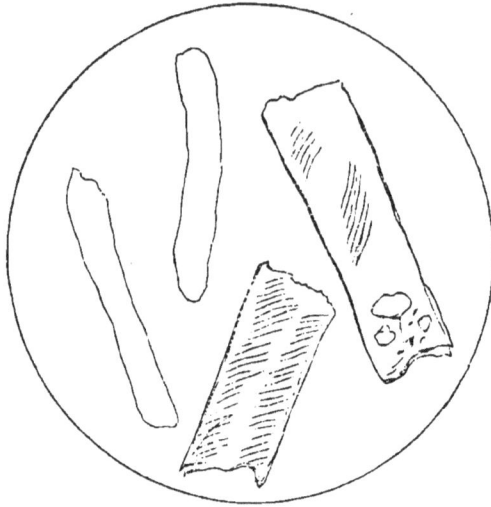

Fig. 8.—Small and large Hyaline Casts composed of pure Fibrine.—× 200.

numerous. When the disease has lasted beyond a month
or six weeks, we find often in adults, more rarely in children, that more or less oil begins to appear in the tube-
casts and in the desquamated renal epithelium. The appearance of oily casts and cells (Fig. 9) excites less alarm
now than it formerly did. It indicates that in certain
parts of the kidney the secreting cells and the inflammatory
exudations are undergoing fatty transformation; but I have
seen many cases of complete recovery after oily casts and
cells in great numbers had appeared in the urine continuously for many weeks.

Morbid Anatomy of the Kidney.—When acute Bright's
disease (acute desquamative nephritis) has proved fatal,
both kidneys are found diseased; they are enlarged and

their weight is increased, each kidney weighing from six to eight ounces, or even more. The increased weight of the kidney is partly due to infiltration with serous fluid ; the tissues contain an excess of water in proportion to solids ; the capsule readily peels off the surface, which is smooth and mottled, presenting an irregular combination of red vascular engorgement, with anæmia and pallor. The fine lobular divisions formed by the minute venous radicles on the surface are more or less obliterated. Some

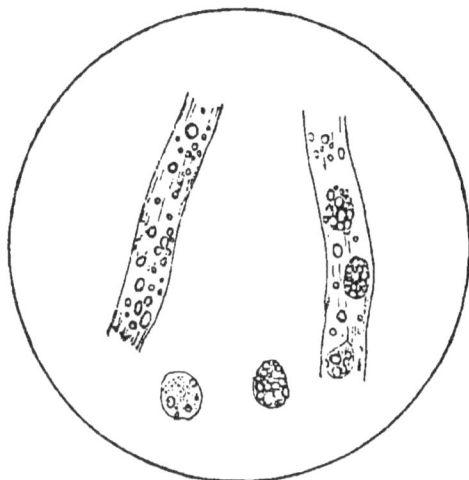

Fig. 9.—Oily Casts and Cells.—× 200.

of the stellate veins are much enlarged and distended. Here and there appear spots of hæmorrhage—some the size and shape of a pin's head, others irregular in form. On section, there appears a marked distinction between the cortical and the medullary portions : the former presents the same mixture of congestion and anæmia as appears on the capsular surface ; the spots of ecchymosis, too, are here visible, but they sometimes take a linear course, especially near the base of a medullary cone. The cones are much darker than the cortex, as if from venous congestion. They appear to be compressed by the swollen portions of the cortex which pass between them, while the bases of the cones are expanded and spread out into the cortical portion, thus having, as Rayer suggested, the form of a wheatsheaf. The mucous membrane of the pelvis of the kidney, and occasionally that of the ureter, is more or less congested.

A microscopic examination of the kidneys shows that the morbid products are contained within the convoluted tubes of the cortex. Most of the tubes are abnormally opaque, in consequence of being filled with epithelial cells which have been formed within them and thrown into their cavity (Fig. 10). The tubes are crowded with these cells

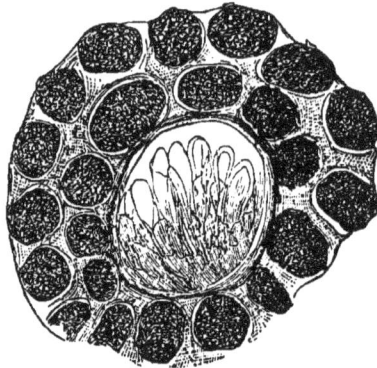

Fig. 10.—Sections and Knuckles of Tubes rendered opaque by their accumulated contents. A Malpighian body near the centre more transparent than the surrounding tubes.— × 200.

in different degrees, some being stuffed full, while in others there is little or no evidence of desquamation having occurred, there being only a single layer of epithelium on their walls, and this being quite normal, or appearing more than usually opaque, granular, and swollen. This condition has been called the ' cloudy swelling ' of the epithelium. The most crowded tubes are usually found in those portions of the cortex which to the naked eye appear pale and anæmic. In these parts the intertubular capillaries are compressed and emptied by the distended and swollen tubes. Occasionally, when examining a section of the tubes, portions of their contents being squeezed out, present exactly the appearance of the 'epithelial casts' which have been before described as existing in the urine.

The hæmorrhagic spots before spoken of, as appearing here and there on the cortical surface and on the face of a section, are seen to be tubes injected with blood which has flowed into them from ruptured Malpighian capillaries (Fig. 11). In some tubes the epithelium is found to con-

Fig. 11.—Malpighian Capsule and Convoluted Tube—the former partially, the latter completely filled with blood from ruptured Malpighian capillaries—thus forming a hæmorrhagic spot in the cortex of the kidney.—×45.

tain more or less oil. Most of the straight tubes of the cones appear to be quite normal, while others are opaque and filled with cells more or less disintegrated, which seem to have been washed into them from the convoluted tubes. There is no evidence that the epithelium of the straight tubes has been morbidly changed.

Some Malpighian bodies are of a deep red colour, with fully injected capillaries; more frequently, however, the Malpighian capillaries appear of a lighter colour than the surrounding opaque tubes; their walls are more opaque than in the normal state, and their surface often appears

rough and finely granular, as if from the coagulation upon
them of some of the fibrinous materials which have trans-
uded through them during life. The nuclei in the walls
of the capillaries are abnormally conspicuous. Occasionally,
as I have before mentioned, there is evidence of rupture of
the capillaries, in the fact that the capsule and the adjoin-
ing tube are filled with extravasated blood (see *ante*,
Fig. 11).

Physiology of the Morbid Process.—The morbid anatomy
of this form of Bright's disease being such as I have described it to be, it remains that we attempt a physiological
explanation of the phenomena. I say a physiological explanation, because I wish to impress upon you that these
morbid phenomena are modifications of normal physiological processes, and admit of explanation only by
reference to physiological principles. The structural
changes in a kidney affected with Bright's disease are the
result of a modified process of secretion. The cortical or
secreting portion of the kidney is obviously the part which
is chiefly implicated, and a microscopical examination
shows that the gland-cells which line the convoluted tubes
are the structures primarily and essentially affected. The
secreting cells of the kidney, like those of other glands,
have the power of separating from the blood and discharg-
ing from the body not only the constituents of their own
proper secretion, but also other materials foreign to that
secretion. Many salts, and many odorous and colouring
matters, when introduced into the circulation through the
stomach, are speedily and completely eliminated through
the kidneys, and apparently without causing structural
change or inconvenience. We do not hesitate to give for
weeks or even months consecutively large doses of such
medicines as iodide of potassium, a salt which is known
to be largely eliminated by the kidneys. It is, however,
important to observe, that certain materials, which when
secreted by the kidneys in moderate quantities for a short

time are quite harmless, may cause decided structural change by their long continued secretion in larger quantities. We have an instance of this in the case of diabetes. Diabetes is not primarily a disease of the kidneys, but kidney-disease is a frequent result of diabetes—indeed, healthy kidneys are rarely if ever found in subjects who have died of diabetes; and the probable reason is, that the long-continued secretion of large quantities of sugar so alters the secreting cells of the kidney, rendering them granular, swollen, opaque, and oily, that at length they lose the power of secreting urine ; the urine becomes albuminous, and complete suppression of the secretion is often the immediate cause of death. Again, when in consequence of obstruction of the gall-duct, or other disease or accident, causing an accumulation of bile in the blood, bile-products in large quantities are secreted by the kidneys, desquamated renal epithelium, tube-casts, and sometimes albumen, are found in the urine. The excretion of these new products sometimes causes a mild form of desquamative nephritis. I refer to these facts to illustrate a physiological principle. It is certain that neither renal gland-cells nor tube-casts are ever found in normal urine, and it is highly probable that the desquamative process never occurs in the kidney except as a result of the excretion of some abnormal materials by the gland. It will scarcely be denied that scarlet fever is associated with a blood-poison. This poison does not always and of necessity implicate the kidneys, as in the vast majority of cases it affects and inflames the skin ; but we know from abundant experience that the risk of renal complication is greatly increased by exposure of a patient to cold while the rash of scarlet fever is out, or even while the skin is desquamating after the disappearance of the rash. It would seem that by exposure to cold the cutaneous inflammation and desquamation are suppressed or checked, and an analogous morbid process is set up in the uriniferous tubes of the kidney. That a poison is thrown off from the skin of a scarlet fever patient, and that the poison is contained in the epidermic scales, we have very good reason to believe; and analogy renders it in the highest degree probable that the implication of the kidneys is associated with the secretion of a morbid poison by their gland-cells. The morbid phenomena result from a modified physiological function. This explanation can scarcely be considered hypothetical; it appears to be the obvious interpretation of unquestionable facts. Meanwhile, the modified cell-growth within the kidney chokes and distends the tubes with desquamated epithelium, the circulation through the gland is impeded, the secretion of urine is checked, and urinary constituents, both liquid and solid, accumulate in the blood. The circulation of urinous blood causes general febrile excitement, with a quick and throbbing pulse, usually a more or less extensive dropsical effusion, and in some cases inflammation of the serous membranes or of other tissues, or serious disorder of the cerebro-spinal functions.

When under favourable circumstances the morbid poison which excited the renal disease has been eliminated, or in part, perhaps, decomposed, the desquamation of epithelium ceases, the secretory process again becomes normal, the urine is copiously secreted, the blood and the tissues are then freed from retained impurities, and from excess of water.

The copious flow of urine which occurs during convalescence from acute Bright's disease is thus explained : during the acute stage of the disease, the constituents of the urine, both solid and liquid, have accumulated in the blood, and have thence been effused into the areolar tissue and into the serous cavities. Now, urea is a most powerful diuretic : when injected into the veins of a dog, it quickly excites an abundant flow of urine ; and as soon as the circulation through the kidney again becomes free, the retained urea exerts its natural diuretic influence upon the gland. The accumulated water serves as a vehicle for washing out the urea, and the copious flow of urine thus induced speedily removes the retained urinary solids and water from the blood, the areolar tissue and the serous cavities into which they had been effused, and thus the dropsy is removed.

This abundant flow of urine, in favourable circumstances, takes place without aid from diuretics or drugs of any kind. I have seen it occur while only a bread-pill or coloured water was given as a *placebo*.

Varieties of Acute Bright's Disease.—I have described to you the usual course of acute Bright's disease associated with a copious desquamation of renal epithelium. For this form of disease I originally proposed the name of *acute desquamative nephritis* ('Med.-Chir. Trans.,' vol. xxx. p. 170). This acute desquamative nephritis is the most common and typical form of acute Bright's disease. But the terms acute Bright's disease and acute desquamative nephritis are not strictly synonymous. There are cases of acute Bright's disease with dropsy which, in all their general features, resemble the cases which I have described as acute desquamative nephritis ; but they differ in this respect, that, from first to last, whether they terminate in recovery or in death, there is no evidence of that process of renal desquamation which forms the characteristic anatomical feature of the cases to which I have before referred. The urine is as scanty and as highly albuminous as in the other class of cases; but it either contains no tube-casts, or it contains, in variable numbers, the small hyaline casts (see Fig. 8), moulded within the clear canal of tubes which retain their lining of gland-cells. When the disease terminates fatally, the kidney presents to the naked eye the same appearances which characterise the acute desquamative cases; but, on microscopic examination, the sections of the convoluted tubes appear very different. The gland-cells are unusually bulky, granular,

and opaque; but the central canal of the tube, instead of
being filled with desquamated epithelium, is clear and
open; so that, while the 'cloudy swelling' of the epithelium renders the margins of the tubes darkly granular
and opaque, the epithelial nuclei being indistinctly seen
or even quite concealed, the central canal of the tube appears comparatively light and clear (Fig. 12).

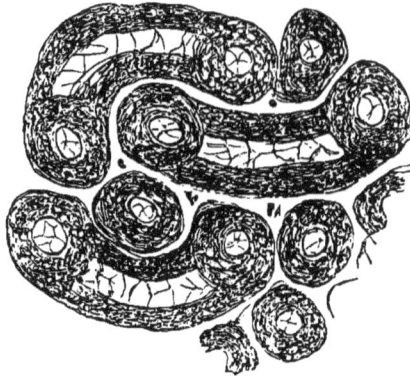

Fig. 12.—Sections of Tubes having Dark Granular Epithelium, with the
Central Canal clear.—× 200.

In other cases of acute Bright's disease, the urine contains few or no epithelial casts; but it deposits a sediment
in which are found numerous casts, mostly of the small
size which indicates that they have been formed within the
central canal of tubes which are lined by gland-cells; and
these small casts contain numerous round cells, which are
identical in appearance with pus-cells and with white
blood-cells (see Fig. 13). I formerly called these 'pus-
casts;' I now call them 'exudation-cell casts.' The name
'pus-cast' is suggestive of suppuration and the formation
of abscess; but no such destructive process is associated
with the appearance of these exudation-cell casts in the
urine. These casts are not unfrequently mixed with epithelial casts in cases of acute Bright's disease; but in
several instances I have seen them in great numbers un-

associated with epithelial casts. I have not been able to discover the exact source of the cells by a microscopic examination of the kidneys; but, since the publication of Cohnheim's researches, it has occurred to me that these exudation-cells may probably be white blood-cells—leucocytes—which have migrated through the walls of the Malpighian capillaries, and subsequently become moulded into small cylindrical casts within the central canal of the convoluted tubes.

In some cases of acute Bright's disease, the exudation-cell casts are of large size (see *c*, Fig. 13). This would

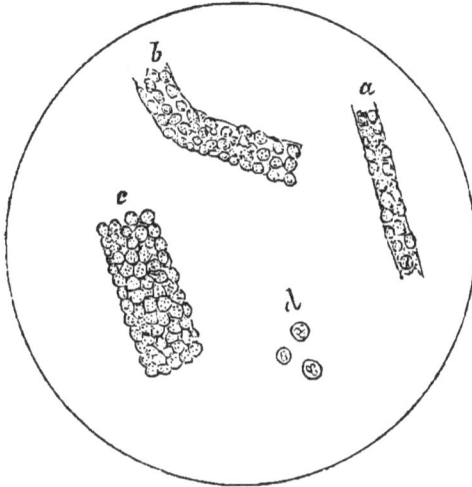

Fig. 13.—Casts entangling small round Exudation-cells, from a case of Acute Bright's Disease. *a* and *b*. Small casts. *c*. A large cast. *d*. Scattered cells.—× 200.

indicate that the cast had occupied the whole diameter of the tube, and so had replaced the epithelial lining of the tube. In some cases, the exudation-cell casts are present from the commencement of the disease; in other cases, the epithelial casts, which were seen at the beginning of an acute attack, are gradually replaced by the exudation-cell casts, which in their turn become mixed with and replaced

by epithelial casts during the progress of recovery and
before the final disappearance of all morbid products from
the urine. I am now describing phenomena which, when
I had ample time and opportunity, I have again and again
observed and noted, and which you may see for yourselves
if you will diligently study the microscopic characters of
the urine from day to day during the progress of acute
Bright's disease. Moreover, I have a permanent record of
the phenomena, in the form of actual specimens of the
tube-casts, which retain their characteristic appearances
after having been kept for years as microscopic objects.

You see, then, that while acute Bright's disease is
usually associated with a more or less copious epithelial
desquamation, there may be no desquamation of renal epithelium, and either no tube-casts or only small hyaline
casts in the urine ; while in other cases, either with or
without epithelial casts, there may be casts crowded with
small exudation-cells. The appearances which I have described are sharply defined in some cases, while in others
they gradually merge into each other. Epithelial casts
and desquamation may be abundant or entirely absent, or
present in moderate amount. The casts with exudation-
cells may be numerous and unassociated with epithelial
casts, or the two forms of tube-casts may be combined with
and replace each other in variable proportions. It is certainly interesting, and, I think, of some practical importance, to note these different appearances in the urine.

In all the cases of acute Bright's disease to which I have
hitherto referred, although the microscopical appearances
in the urine are various, the general symptoms and the
physical and chemical characters of the secretion are alike,
and in particular the presence of a large amount of albumen is a constant phenomenon. Now, I have to tell you
that we sometimes, though rarely, meet with cases of acute
general dropsy in which the urine, although scanty, contains not a trace of albumen. In the great majority of

cases, acute Bright's disease and acute albuminuria are synonymous terms; but in these few exceptional cases the latter term is inapplicable, for the urine is not albuminous. Dr. Blackall described two cases of acute general dropsy after scarlet fever in which the urine was not coagulable either by heat or by nitric, or, as he calls it, 'nitrous' acid (*op. cit.* pp. 12 to 21). Dr. Roberts gives the history of two cases after scarlet fever, both fatal—one acute, the other chronic ('On Urinary and Renal Diseases,' pp. 24 and 400). Dr. Basham has recorded the case of an adult in whom general dropsy followed exposure to wet and cold. He recovered ('Lancet' August 1867.) Dr. Dickinson has published one fatal case in a child eighteen months old ('On the Pathology and Treatment of Albuminuria,' p. 73.) And I have notes of four cases that have come under my own observation. Three of these cases recovered, and the fourth was improving when he was lost sight of. In two of my cases, the dropsy followed scarlet fever, and in the other two it was probably a result of exposure to cold. In two of the cases, neither albumen nor tube-casts could be discovered throughout; in one, a trace of albumen was found on one occasion; and in the fourth, after general dropsy had existed for six weeks without albumen or tube-casts, a trace of albumen and some hyaline casts appeared.

Now what is the explanation of these rare, remarkable, and exceptional cases? I have neither seen nor heard of any satisfactory explanation of them, and I am not prepared to give you one; but I venture upon one or two suggestions and queries. There is reason to believe that suppressed action of the skin is a powerful concurring cause of the dropsy which is associated with albuminuria; and this, perhaps, is the explanation of the frequent association of dropsy with the renal disease which results from scarlet fever or from exposure to cold and wet. In both these classes of cases, the functions of the skin must obviously be more or less impaired—in the one by the

specific inflammation, and in the other by cold; whereas
diphtheritic albuminuria, without implication of the skin
in the morbid process, rarely if ever gives rise to general
dropsy. Then the question arises, Is it possible that suppression of the cutaneous secretion may alone cause acute
general dropsy without the implication of the kidneys? May
acute general dropsy result from a metastasis of the perspiration from the skin to the areolar tissue and the serous
membranes? And may the scanty secretion of urine in these
exceptional cases be a result of the morbid transfer of
water to the tissues where the dropsical effusion takes
place, as, by a reversed action, the perspiration is checked,
and the skin of a diabetic patient rendered dry by the
copious flux of liquid through the kidneys? I am not
prepared to answer these questions. In most cases of
acute dropsy without albuminuria, the urine has been
scanty and high coloured. In one of Dr. Roberts' cases,
the urine was scanty almost to suppression, only two
drachms having been voided in twenty-four hours; 'it
contained casts, but not a trace of albumen.' The form
of tube-casts is not mentioned. The total quantity of
urine voided during the last seven days of life amounted
to between six and seven ounces. No autopsy was permitted. In the second case, the urine contained neither
albumen nor casts : but it was scanty and high coloured;
and, death having occurred after an illness of five months,
'the kidneys were found to be good examples of the
smooth white Bright's kidney.' It this case it would seem
that, although there was no albuminuria, there was some
structural change in the kidneys. Dr. Wilks has published
in the sixth volume of the 'Pathological Transactions' a
remarkable case of general dropsy, with a peculiar form of
renal disease, but without albuminuria, in a woman aged
35. The urine passed amounted to about twelve ounces in the
day, of specific gravity 1012, and not albuminous. A few
days before death, the urine became less in quantity, and
for the last four days none was obtained. The kidneys

were pale and large, their combined weight being seventeen ounces. The cortical portion was seen by the naked eye to be scattered over with small round dots like grains of sand. On a microscopic examination, these were found to be the Malpighian bodies, the capillaries of which were covered over with mulberry-like masses of oil-globules, while the tubes were healthy.

It may hereafter happen to some of you to have the opportunity of throwing additional light upon the pathology of these rare and exceptional cases of general dropsy not dependent on heart disease and unassociated with albuminuria.

We have seen that the chief varieties and modifications of acute Bright's disease with albuminuria are the following:—1, with epithelial desquamation (desquamative nephritis); 2, without desquamation, either with or without small hyaline casts; 3, with exudation-cell casts, either with or without epithelial casts and desquamation. Lastly, we have, as an entirely distinct class of cases, rare, exceptional, and obscure in their pathology, acute Bright's disease, or at any rate acute general and febrile dropsy, without albuminuria.

Changes in the Blood.—The effect of acute Bright's disease is not only to cause an admixture of blood-constituents with the urine, but also to bring about a large accumulation of urinary materials in the blood. While the urine is usually more or less bloody, the blood becomes in a greater or less degree urinous. Dr. Christison was the first to announce the fact that the blood in these cases contains a large amount of urea, and that urea is found in the dropsical and inflammatory effusions ('Edinburgh Medical and Surgical Journal,' October 1829). Not only is the blood altered by an accumulation of urinary materials, but also by a loss of its own normal constituents. The density of the serum is reduced from 1030 or 1031 to 1022 or even 1020. The loss of density is greatest when the urine has been most albuminous; and it is probably

explained by the escape of serum through the kidneys. The
hæmoglobin or colouring matter also diminishes rapidly,
the normal proportion being 1,335 in 10,000, Dr. Christison found it reduced, after a few weeks' illness, as low as
955 in one case, in another to 564 ; and in a young man
ill for three months and a half subsequent to scarlet fever,
who had never been bled before, it was only 427. At the
commencement of the disease, the loss of colouring matter
is less rapid than the extreme pallor of the patient would
seem to indicate ; and it is probable that the blanched
appearance of the skin is partly occasioned by the quantity of water in the subcutaneous tissue.

Etiology.—Acute Bright's disease may occur at all ages
from infancy to extreme old age. The two most frequent
causes of acute Bright's disease with dropsy are exposure
to wet and cold and scarlet fever. Either of these causes
is alone sufficient to excite the disease ; but their combined action—exposure to cold during the progress of
scarlet fever—is a most powerful determining cause of the
malady. Diphtheria is a frequent cause of albuminuria ;
but, as I have before said, diphtheritic albuminuria is
rarely associated with dropsy. Amongst the less frequent
causes of acute albuminuria are measles, erysipelas, pyæmia,
the absorption of poisonous materials from the uterus after
parturition, rheumatic fever, the malarious poison, typhus
and typhoid fever, cholera, and, lastly, excessive eating and
drinking, more especially when combined with dyspepsia.
The chain of events which connects albuminuria with dyspepsia is probably this—imperfectly digested food passes
into the blood and loads it with impurities. The gland-
cells of the kidney excrete these ill-digested products, and
in doing so undergo structural changes, while the imperfectly assimilated albuminous materials pass more readily
by exosmosis through the Malpighian capillaries. Further,
the malnutrition resulting from chronic dyspepsia causes a
general nervous exhaustion, with loss of vaso-motor-nerve

force, and consequent diminution of tone and contractile power in the muscular walls of the minute arteries generally, including those of the kidney, while the walls of the capillaries are probably weakened by depraved nutrition. Thus the filter and the fluid to be filtered are both materially changed, while the increasing impurity of the blood throws more work upon the kidneys, and favours the passage of the altered albumen through the kidneys, which is often notably increased after food.

In the majority of cases, acute albuminuria resulting from other causes than scarlet fever and exposure to cold is unassociated with dropsy, and its history belongs to that of the disease with which it is associated as a complication. We shall find hereafter that albuminuria resulting from one or other of the various causes here referred to sometimes leads to a chronic and incurable degeneration of the kidney. Excess of alcohol is a more frequent cause of chronic than of acute Bright's disease. A remarkable case of transient alcoholic albuminuria occurred when Dr. Baxter was house-physician to our hospital. A man between twenty and thirty years of age was brought in one night by the police. He was unconscious, and breathing stertorously. He was believed to be drunk, and a large quantity of vinous liquid was pumped out of his stomach. The unconsciousness remaining, uræmia was suspected, and some urine drawn off with the catheter was ' loaded with albumen.' He was then put into bed, cupped over the loins, and a purgative given. When Dr. Baxter visited the ward in the morning, he found the man sitting up and clamouring for his discharge. He said that he had been very drunk over-night, but now he had nothing the matter with him. He passed some urine, which was found of normal colour and specific gravity, and without a trace of albumen. He then left the hospital in triumph. The temporary albuminuria was the result of renal congestion while the excess of alcohol was being excreted by the kidneys.

Diagnosis.—In most cases of acute Bright's disease, the symptoms are so obvious that the disease can scarcely be overlooked or mistaken for any other. The only cases in which there is a possibility of acute albuminuria being unrecognised are those in which it is unassociated with dropsy. But, the existence of albuminuria being discovered, it is not always easy to determine whether this is the result of a recent acute attack, or of a chronic degeneration of the kidney. We shall be in a better position to discuss this important practical question after we have studied the various forms of chronic Bright's disease. Meanwhile, however, I may tell you that, as a rule, high coloured, smoky, and blood-tinged urine, of high specific gravity, is an indication of a recent acute attack; and equally so is a copious sediment composed of epithelial and blood-casts (Figs. 6 and 7), or of exudation-cell casts (Fig. 13), alone or mixed with epithelial casts. The appearance of oily casts and cells (Fig. 9), in combination with numerous epithelial casts, does not materially affect the diagnosis. On the other hand, urine of low specific gravity and very pale in colour, yet highly albuminous, is usually evidence of chronic disease; and this evidence is strengthened by the appearance of numerous oily casts and cells unassociated with epithelial or exudation-casts. Large hyaline casts (Fig. 8) in *pale* highly albuminous urine point to disease not only chronic, but in an advanced stage. We shall return to this subject, and discuss it more fully in a future lecture.

Prognosis.—Acute Bright's disease has a tendency to terminate in complete recovery. It is essentially a curable disease, as much so as acute bronchitis or acute pneumonia. The earlier the patient comes under treatment, the better is his prospect of recovery; and, on the other hand, the longer the symptoms have continued without signs of amendment, the more grave does the prognosis become. The prognosis is, on the whole, more favourable in the young and middle-aged than in those more advanced in years;

but the disease may prove mild and tractable even in very aged persons. For obvious reasons, the prospect of recovery is better in the case of those who can avoid exposure to cold and other injurious influences, than when the patient's circumstances are less favourable.

In favourable cases, a copious secretion of urine, of comparatively low specific gravity and of paler colour, with a diminishing amount of albumen and decrease of dropsy, are amongst the earliest signs of amendment. Albuminuria is usually the last symptom to disappear. The time of its disappearance varies, in different cases of recovery, from a few days to many. months. If the urine continue albuminous for more than six months, it becomes more and more doubtful whether it will ever cease to be so; but I have seen cases of complete recovery after albuminuria had continued for one, two, or even three years. So long as the urine continues albuminous, in however slight a degree, although the dropsy and all other general symptoms may have passed away, recovery must be considered incomplete. Acute Bright's disease, although, as a rule, a curable, is not unfrequently a fatal disease. There are some symptoms and complications which indicate a case of more than ordinary peril; such as a very scanty secretion of highly albuminous urine; frequent and distressing vomiting; great anasarca, with a tendency to erysipelatous inflammation of the skin; dropsical effusion within the chest, either in the pleura or the pericardium, or both, with urgent dyspnœa; inflammation of the lung, or pleura, or pericardium, or endocardium; severe and persistent headache, which is apt to be followed by convulsions and by coma, with a brown and dry tongue. All these are symptoms of grave, though not always of fatal, import. When the renal disease is acute, and therefore essentially curable, recovery sometimes occurs after the most formidable symptoms of uræmic poisoning have been present.

A consideration of the exciting causes of the renal

disease forms an element in the prognostic indications. When Bright's disease results from some inherent constitutional defect, without obvious exciting cause, it is generally more intractable than when it is directly due to exposure to cold or to the influence of some specific blood-poison, as, for instance, that of scarlet fever or erysipelas. To all general rules of this kind there are exceptions, and each case requires a separate and careful study.

Let me impress upon you one point of practical importance. Before you pronounce a patient to be entirely free from his malady, be careful to test the urine, not only after rest and fasting—*i. e.*, in the morning before breakfast —but after food and exercise. Albuminous urine is usually more copiously so after food and exercise ; and you will sometimes find that, while the urine before breakfast is quite free from albumen, that which is secreted after a meal is decidedly and even copiously albuminous. In some cases, exercise has even more influence than food in exciting renal congestion and albuminuria.

An attack of acute Bright's disease confers no immunity from future attacks; on the contrary, the disease may occur more than once in the same subject, a result either of inexplicable predisposition or of a liability resulting from a first attack. I think my experience warrants the statement that when acute albuminuria has resulted from some non-specific cause, such as exposure to cold and wet or excessive eating and drinking, it is more likely to recur than when it has been excited by a specific morbid poison, such as that of scarlet fever, which, as a rule, does not occur a second time in the same individual ; but I have known patients so unfortunate as to have two attacks of scarlet fever, and each attack complicated with albuminuria. I shall defer the important question of treatment until we have passed in review the various forms and complications of chronic Bright's disease.

LECTURE III.

CHRONIC BRIGHT'S DISEASE.

Small Red Granular Kidney—Synonyms—Outward Appearance of the Kidney in different Stages—General History of the Disease—Chemical and Microscopical Characters of the Urine—Microscopic Appearances in the Kidney—The Structural Changes are essentially tubular and intra-tubular—Changes in the Blood-vessels of the Kidney—Physiological Explanation of the Structural Changes in the Kidney and of the Condition of the Urine—Local and General Symptoms of Contracted Granular Kidney; Frequent Micturition—Pain in the Back—Dyspepsia as a Cause and a Consequence of Renal Disease—Dropsy—Hypertrophy of the Heart—Inflammation of Serous Membranes—Hæmorrhage from various Surfaces—Cerebral Hæmorrhage—Impairment of Vision—Cerebral Symptoms the Result of Uræmia—Theory of Uræmia—Nervous Dyspnœa—Disease of the Liver—Diagnosis—Prognosis.

Classification of Chronic Bright's Disease.—In my last lecture I gave you some account of acute Bright's disease, and I now proceed to discuss the subject of chronic Bright's disease. Cases of chronic Bright's disease arrange themselves, anatomically and clinically, in two very distinct classes. In one class of cases, the kidney is found small, red, and granular; in the other class, on the contrary, the kidney is large, pale, and usually smooth on the surface.

Small Red Granular Kidney.—The clinical history of the two classes of cases is as distinct as are their anatomical characters. For various reasons, it will be more convenient to take first in order those cases which are associated with the small red granular kidney. In the Nomenclature of the Royal College of Physicians the disease is designated ' granular kidney,' with the *synonyms* ' contracted granular kidney,' ' chronic desquamative nephritis,' ' gouty kidney.'

Outward Appearance of the Kidney in different stages.

—I place before you drawings representing kidneys in different stages of degeneration. At no period of the disease is there enlargement of the kidney, but from the commencement a process of wasting occurs. In the early stage, when death has occurred from some other disease, the capsule is found adhering firmly to the surface of the gland, so that it is difficult to tear it off without bringing away some of the adherent glandular tissue. The fine lobular markings are less distinct than in the normal state, and the surface of the kidney is slightly uneven and granular. As the disease advances, there is progressive wasting of the glandular portion of the kidney, with granular unevenness of the surface and diminution of the thickness of the cortex; so that by degrees the bases of the medullary cones approach nearer to the surface of the gland. In extreme cases, the kidney may be reduced to one-half or even one-third of its normal size and weight. The cortical secreting portion of the gland is evidently the part chiefly implicated, while the medullary cones are nearly intact. The contracted kidney is somewhat firmer and tougher than natural. In all stages of the disease, one or more, sometimes numerous, serous cysts may be seen projecting from the surface, and varying in size from a pin's head to a pea, but sometimes as large as a filbert, or even larger. Even in the most advanced stages of atrophy, the organ retains more or less of its normal red colour and its vascularity ; hence it is called the *red* granular kidney, to distinguish it from certain cases of chronic Bright's disease to be referred to hereafter, in which the kidneys are granular, but white and anæmic.

General History of the Disease.—Some general facts relating to the disease it may be well to point out now. The disease is essentially chronic from the commencement, and rarely, if ever, a sequel of an acute attack. Its commencement, therefore, is, as a rule, insidious, and in its early stages it is often unsuspected and latent. It is a comparatively rare disease in early life, not uncommon between the ages of twenty and thirty ; but the majority of cases occur in persons at and beyond middle age. It is often associated with the gouty diathesis, as one of its synonyms indicates; and it is of common occurrence in persons who eat and drink to excess, or who, not being intemperate in food or drink, suffer from certain forms of dyspepsia, without the complication of gouty paroxysms. In some cases, the disease probably results from habitual exposure to cold and wet, and consequent suppression of the cutaneous secretion. There is reason to believe that chronic poisoning by lead is, at any rate, a concurring cause of the disease amongst painters and others who are exposed to the influence of this pernicious metal. Dr. Garrod was the first to direct attention to the influence of lead in the causation of gout; and Dr. Dickinson states that, out of forty-two men exposed to lead-poisoning who had died in St. George's Hospital, twenty-six had granular degeneration of the kidneys, which in most cases was so advanced as to have caused death ('On the Pathology and Treatment of Albuminuria'). Allowing, as we must, that the lead had great influence, it is probable that habits of intemperance and other causes may have co-operated with the lead. Granular kidney is occasionally, though rarely, found as a sequence of the albuminuria which is associated with pregnancy. I have seen one well-marked instance of this. The atrophy with granulation which results from passive congestion of the kidney consequent on valvular disease of the heart or emphysema with chronic bronchitis, has a different pathological history ; and I shall refer to it on a future occasion.

During the progress of the disease which results in the contracted granular kidney, dropsy rarely forms a prominent symptom, and in the majority of cases it is entirely absent. The disease is often associated with hypertrophy of the left ventricle of the heart, even when there is no valvular defect or disease of the walls of the larger arteries to explain the cardiac hypertrophy. In a large proportion of cases, the immediate cause of death is uræmia or cerebral hæmorrhage.

Now, in the course of these lectures, I shall as much as possible avoid all controversial topics ; but, in proceeding to give you what I believe to be the true account of the minute anatomy and pathology of this disease, I am bound to tell you that I dissent from the opinions of some pathologists for whom I entertain great respect, but not sufficient to induce me to follow them into what I believe to be an erroneous reading and interpretation of facts. Virchow, in his ' Cellular Pathology,' states that there are three anatomical elements in the kidney—namely, tubes, vessels, and interstitial tissue ; and, in accordance with this, there are three forms of Bright's disease—what he calls parenchymatous nephritis, having its seat in the tubes; amyloid degeneration in the blood-vessels; and interstitial nephritis, consisting essentially, as he believes, in thickening of intertubular tissue and consequent atrophy and granular contraction of the kidney. Virchow admits that two and sometimes all three of his forms of disease may coexist in the same kidney ; and I maintain that in every case of Bright's disease all the tissues are implicated; the various forms of disease depending, not upon the implication of different anatomical elements in the morbid process, but upon the varying nature of the structural changes which these elements undergo in different classes of cases. I will endeavour to make this clear as I proceed. My doctrine with regard to the minute anatomy and pathology of the granular kidney is, that it consists primarily and essentially in a disintegration and destruction of the gland-cells which line the convoluted tubes, the *débris* of the gland-cells appearing in the urine as granular tube-casts ; that the destruction of the gland-cells induces atrophy and contraction of the tubes; that this shrinking of the tubes, with some thickening of their membranous walls and of the Malpighian capsules, gives a delusive appearance of interstitial or intertubular formation of fibrous tissue; and that thickening of the walls of the arteries, the nature of which I shall presently describe, forms one of the most constant and conspicuous features of the disease ; although this arterial change is entirely ignored by Virchow and his followers, who erroneously assume that the so-called amyloid or waxy degeneration is the only form of Bright's disease constantly and essentially associated with thickening of the blood-vessels.

Chemical and Microscopical Characters of the Urine. —You will find that the minute structural changes in the contracted kidney are rendered easily intelligible if you study them in connexion with the clinical history of the disease, and in particular with the chemical and microscopical characters of the urine. I have told you that the disease, although not exclusively of gouty origin, is often associated with chronic gout. Examine the urine of a man who has had repeated attacks of gout, and you will not unfrequently find in it the earliest indications of incipient renal degeneration. The urine may be of normal colour and specific gravity, and without a trace of albumen ; but, after standing for a few hours in a conical glass, it deposits a light cloud, which, on microscopic examination, is found to consist of scattered granular *débris* and tube-casts such as are represented in Fig. 14. These casts contain epithelial cells in various grades of disintegration, and hence arises their 'granular' appearance. Every granular cast is not of necessity composed of disintegrated epithelium. Blood-corpuscles may become disintegrated within the uriniferous tubes, and appear in the urine as granular blood-casts, distinguished from granular epithelial casts by their reddish-brown colour, and often by containing some entire blood-corpuscles ; so, disintegrated hyaline casts may assume a granular appearance ;

E

but by a comparison with other casts associated with them,
and by noting the various grades of change, we trace them
to their true source. The presence of granular epithelial
casts and of scattered epithelial *débris* is evidence that a
process of epithelial desquamation and disintegration has
commenced in the kidney.   In the earlier stages of the
renal disease, the granular casts are found only during or

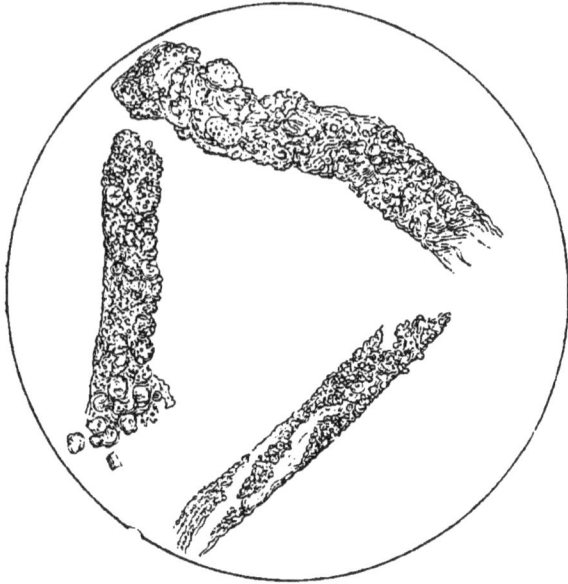

Fig. 14.—Granular Casts, composed of more or less completely Disintegrated
Epithelium and Fibrine.— × 200.

immediately after a gouty paroxysm, and, as I have
already said, unassociated with albuminuria. In the intervals between the attacks of gout, no tube-casts are visible.
At a later stage, granular casts and epithelial *débris* are
always present in a greater or less amount ; and the urine
becomes albuminous, at first only during a fit of gout, the
tube-casts also being more abundant during the paroxysm.
At a still later period, tube-casts and albumen are more or
less constantly present, though both may be absent even

in the most advanced stages of this form of disease. Not only is the particular form of renal disease indicated by the microscopic appearance of the urinary sediment, but the number of granular tube-casts and the amount of epithelial *débris* indicate the rate at which the disease is progressing. The more copious the sediment, the more rapid is the destruction of the gland-cells, and the consequent atrophy of the kidney. In the more advanced stages of the disease, large hyaline casts are often found associated with the granular casts. (Fig. 15.) As the

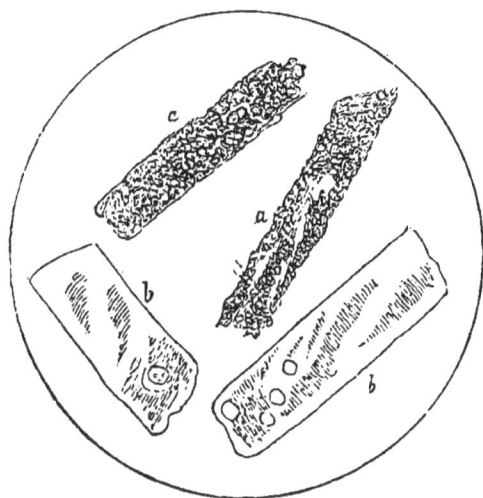

Fig. 15.—*a a.* Granular Casts. *b b.* Large Hyaline Casts.—× 200.

disease makes progress, the urine undergoes remarkable physical changes. The quantity secreted is usually in excess of the normal amount; and with the increase of quantity there is commonly associated a loss of the natural colour and a diminution of the specific gravity, which, usually as low as 1010 or 1012, sometimes falls to 1005. The low specific gravity indicates a relative decrease of the normal solid constituents, especially of urea, uric acid, and extractive matters. In one of Dr. Christison's cases, the total solids discharged were reduced to one-fifth, and in another nearly to one-twelfth, of the healthy average. This defective discharge of solids is partly explained by the rest in bed, the scanty diet, and the general anæmia. The amount of albumen varies considerably. Absent or scanty in the early stage, it may be rather copious in the middle periods, and again scanty or even entirely absent in the stage of extreme degeneration of the kidney.

Microscopic Appearances in the Kidney.—The kidneys should be examined as soon as possible after death, and before the appearance of their tissues has become changed by any antiseptic or hardening process. Thin sections of the cortex, made with a Valentin's knife, may be placed in a solution of common salt and water of specific gravity 1030, and then examined with a magnifying power of not less than 200 diameters. Dilute acetic acid brings out some of the appearances very distinctly. The chief changes

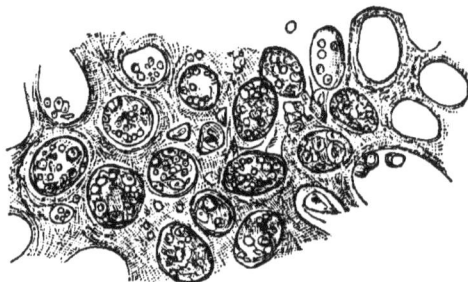

Fig. 16.—Transverse Sections of Tubes, containing only Granular *Débris* of Epithelium held in position by Coagulated Fibrine. At one end of the section, the contents of the Tubes have been washed away, and the sections of the basement-membrane form three empty rings.— × 200.

will be found in the convoluted tubes, in the arteries, and in the Malpighian capsules and capillaries. In some tubes the gland-cells have their normal appearance, or they are opaque and granular, with a clear central canal (see *ante*, Fig. 12). Other tubes are filled and rendered opaque with desquamated epithelium more or less disintegrated (see *ante*, Fig. 10). Others, again, present the characteristic ap-

pearance represented in Fig. 16. Their epithelial lining
has become disintegrated and removed, appearing in the
urine in the form of the granular casts before described
(Figs. 14 and 15). A few granular particles of epithelium
only remain, and these appear to be held together by
fibrinous coagula. The transverse sections of tubes in this
condition have somewhat the appearance of oval or globular cysts, and many years ago they were described as
microscopic cells by a very able observer (see Mr. Simon's
paper on Subacute Inflammation of the Kidney, 'Med.-
Chir. Trans.,' vol. xxx). When, in the same section, some
tubes appear transversely divided, while others present
themselves lengthwise, as in Fig. 17, all having the same

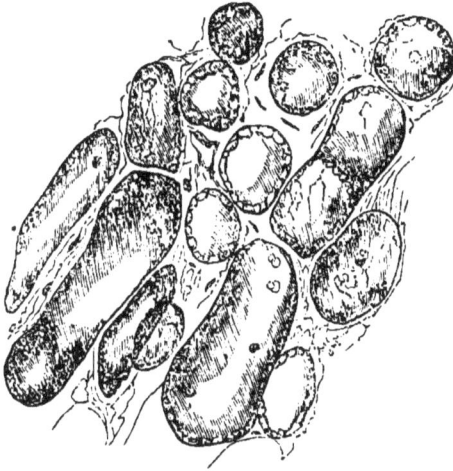

Fig. 17.—Tubes more or less completely denuded of Epithelium. Some
transversely divided and cyst-like ; others seen lengthwise.—× 200.

general structure and contents, it is easily seen that the
cyst-like appearance is given by transverse sections of partially or completely denuded tubes.

The number of tubes thus denuded of their epithelial
lining varies much in different cases. In some kidneys,
which to the naked eye present comparatively little devia-
tion from the normal state, the destruction of gland-cells

is found to have been very extensive. Other tubes are found, as regards their appearance and contents, in the same condition as those just now described, but apparently shrunken and atrophied, with wide interspaces between them—the interspaces being occupied by the remains of other atrophied tubes and capillaries (see Fig. 18). Atrophy of the tubes appears to be the usual result of the destruc--

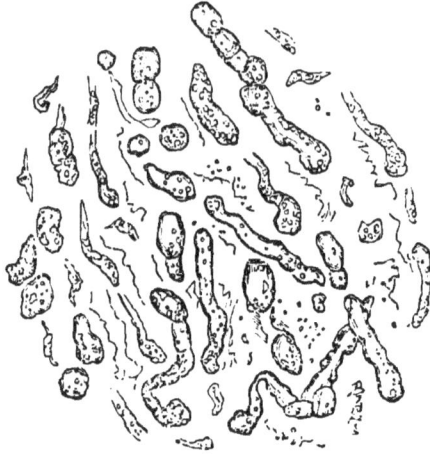

Fig. 18.—Tubes in process of Atrophy and Contraction after the Destruction of their Epithelial Lining, a few granular particles only remaining within them.—× 200.

tion and removal of their gland-cells. But the opposite condition of dilatation is found in some of the tubes, which may be seen often as large as Malpighian bodies, and even larger (Fig. 19); and there can be no doubt that these dilating tubes at length form the cysts which are visible by the naked eye.

There is yet another appearance of the tubes in the granular kidney, which probably has a close relation with the formation of dilated tubes and cysts. This appearance is represented in Fig. 20. The tubes are lined by layers of delicate clear-walled cells of a more or less rounded form, and each having a single nucleus. Some tubes in this condition may be found in every granular kidney, but their number varies considerably. In most contracted kidneys, these round-celled tubes are relatively few in number; in others they are very numerous; and I believe there is a special relationship between these tubes and the

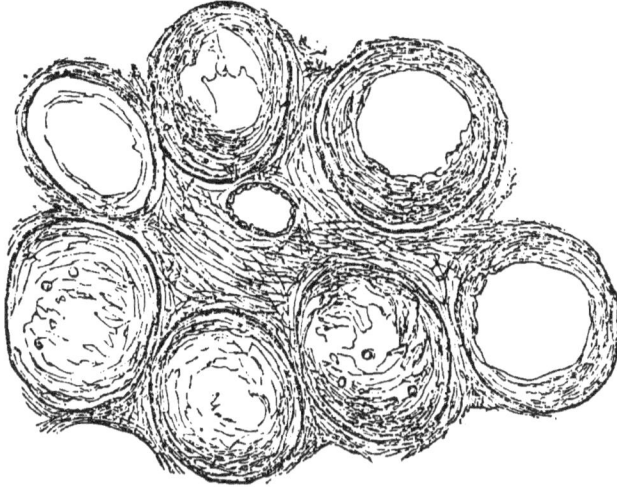

Fig. 19.—Transverse Sections of Dilated Tubes with Thickened Walls. In some sections, the open mouth of the divided tube is seen—one section of a denuded tube of normal size.— ×200.

cysts with which they are usually associated. It can scarcely be doubted that renal cysts are dilated tubes; and

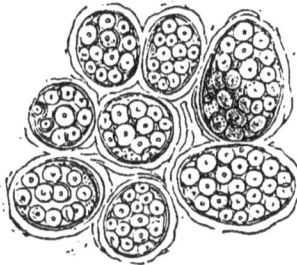

Fig. 20.—Sections of Tubes, in which a layer of Transparent Cells, each with a single Nucleus, has taken the place of the Normal Epithelial Lining.— ×200.

it is probable that the tubes are dilated by the accumulation of a watery fluid secreted by these delicate cells, which I have described. This at any rate is certain, that the contracted granular kidney is the only form of Bright's disease with which, as a rule, cysts are associated ; and it is in these kidneys only that we find tubes lined by the transparent cells in question. (See a remarkable case recorded by Dr. Conway Evans, ' Pathological Transactions,' vol. v. p. 183. I examined the kidneys in that case, and can confirm the accuracy of Dr. Evans's description.) It is probable that the dilatation of the tubes, by their accumulated liquid contents, and their conversion into aqueous cysts, is favoured by the obstruction of their outlets through an accumulation and impaction of epithelial and fibrinous *débris.*

I have yet to mention that here and there in the cortex of the kidney a tube may be seen without gland-cells, and completely filled with disorganised fibrine, which, if it had escaped during life, would have appeared in the urine as a large hyaline cast (see Fig. 15). Sometimes, though rarely in this form of disease, a tube is found injected with blood which has escaped from ruptured Malpighian capillaries. In some few of the tubes the contents have undergone a fatty transformation ; and oil-globules may be seen either contained within cells, or free and irregularly accumulated within the tubes.

The basement-membrane of the tubes usually appears to be somewhat thickened ; and this thickening, together with the wide spaces between the tubes occupied by the atrophied remains of shrunken tubes and intertubular capillaries (see Fig. 18), has given rise to the doctrine of an excess or of a new formation of connective tissue between the tubes. Now, as I told you in my first lecture, there is in the normal condition no organised connective tissue between the convoluted tubes. In this statement I am in accord with Ludwig, who says, ' No fibrillated connective tissue exists between the tortuous portions of the urinary

tubules ' (see Stricker's ' Human and Comparative Histology, New Sydenham Society's Translations,' vol. ii. p. 106);
and I show you thin sections of beautifully injected healthy
kidneys, in which you can see that the intertubular capillaries are in immediate contact with the outer surface of
the basement-membrane, and that if anything intervenes,
it is quite homogeneous, gelatinous, and structureless. In
fact, the basement-membrane of the tubes constitutes the
true and only connective tissue between the intertubular
capillaries and the intratubular epithelium ; therefore,
whatever increase of this tissue there is—and it is much
less in the granular kidney than in some large white or
lardaceous kidneys—is not an intertubular, but a tubular,
thickening. It is obvious, however, that if we take into
account the wasted tubes which remain in a shrunken
condition after the destruction of their epithelial contents,
there is in the granular kidney a relative excess of basement-membrane connective tissue.

There is often an appearance of fibrous connective tissue
round the Malpighian bodies. I am not prepared to say
that no connective tissue is ever formed external to and
apart from the fibrous capsule of the Malpighian body ;
but I am convinced that, when the capsule is somewhat
thickened, as it often is in the granular kidney, and then
thrown into folds, the fibrous appearance which it presents,
as different depths of the globular capsule are being focused
into view, may very readily be mistaken for fibrous connective tissue outside and surrounding the capsule (see
Fig. 21).

In consequence of wasting and contraction of the tubes,
some of the Malpighian bodies are brought nearer together,
and three or four may sometimes be seen almost in contact with each other. Sections of dilated tubes, such as
are represented in Fig. 19, may easily be mistaken for
Malpighian bodies ; but in the sections of tubes the open
mouth of the cut tube may often be seen, and in the Mal-

pighian bodies the capillaries, with the nuclei in their walls, are characteristic.

Now, to recapitulate. We have found the following pathological appearances in the tubes : the epithelium opaque and granular, in a state of cloudy swelling ; the tubes crowded and opaque, with degenerated and disintegrated epithelium ; some tubes deprived of their epithelium ;

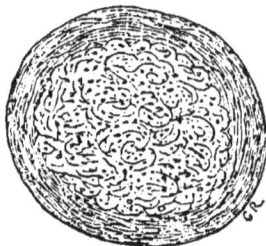

Fig. 21.—Malpighian Body—the Capsule thickened and having a Fibrous Appearance. The capillaries thickened and opaque ; the nuclei visible in their walls.— × 200.

some contracted ; others dilated in various degrees ; some lined by transparent uninucleated cells ; others filled with unorganised fibrine, rarely with blood, or with oil ; lastly, the basement-membrane and the Malpighian capsules thickened, this thickening being often more apparent than real. All the changes which I have described, and which you may see in the specimens which I have placed under the microscopes on the table, are essentially tubular and intratubular. You see, then, with how little reason the contracted kidney is spoken of as the result of an essentially intertubular disease.

Changes in the Blood-vessels of the Kidney.—Amongst the most constant and interesting anatomical changes are those which occur in the minute arteries. The walls of the minute renal arteries present in the normal state two layers of fibres, an inner longitudinal and an outer circular layer (see Fig. 22). In the advanced stages of contracted granular kidney, both these layers are much hypertrophied.

The two layers remain quite distinct, and sharply defined.
There is an excess of normal muscular tissue with no ap-

Fig. 22.—Normal Artery from the Kidney.— × 200.

pearance of structural change or degeneration (see Fig. 23).
The hypertrophy is usually most conspicuous in the smallest
arterial branches, and a comparison of the afferent artery

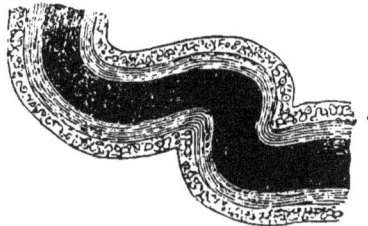

Fig. 23.—Artery, with Hypertrophied Muscular Walls, from the Kidney. An inner
longitudinal and an outer circular layer of fibres, of about equal thickness. The
canal is injected.— × 200.

of the Malpighian body in a healthy kidney with one from
a small granular kidney will often show that the arterial
walls in the latter are twice or even thrice the normal
thickness.

A transverse section of an artery of larger size, such as
is represented in Fig. 24, shows the projecting cut ends
of the longitudinal layer of hypertrophied muscular fibres
surrounded by the outer circular layer. You may see for
yourselves that, in the specimens under the microscopes,
the appearances described are as distinctly visible as in the
diagram.

I discovered this remarkable hypertrophy of the muscular
walls of the renal arteries more than twenty years ago, and
published the fact in the 'Medico-Chirurgical Transactions,' vol. xxxiii. Since that time, I have found arterial
hypertrophy in every contracted granular kidney that I

have examined. It is not present in the earlier stages of the disease, but it comes on gradually, and proceeds *pari passu* with the structural changes within the uriniferous tubes. This thickening of the arterial walls from overgrowth of their normal muscular tissue is quite distinct from that structural change which I shall hereafter show you, and describe as the waxy or lardaceous degeneration of the blood-vessels.

The walls of the Malpighian capillaries are usually thickened and opaque, their surface sometimes smooth and

Fig. 24.—Transverse Section of a Renal Artery, with Hypertrophied Muscular Walls. The inner longitudinal and the outer circular layer of fibres clearly seen.— × 200.

wax-like, more commonly finely granular (Fig. 21). The thickening of the capillary walls, together with that of the Malpighian capsule, tends to conceal the blood within the Malpighian capillaries, and gives the Malpighian bodies a dull grey appearance. The intertubular capillaries, and the veins into which they empty themselves, present no appearance of thickening.

Physiological Explanation of the Structural Changes in the Kidney.—In a paper published long since (' Med.- Chir. Trans.' vol. xxx.) I designated the disease which I am now describing *chronic desquamative nephritis.* The term ' chronic desquamative *disease*' is, I now think, preferable, since it implies no theory as to the inflammatory nature of the disease. The primary and essential structural changes consist in a desquamation, disintegration, and removal of the renal gland-cells; and the pathological process admits of the same physiological explanation as that which, in my last lecture, I gave of the acute desquamative disease. The changes in the glandular epithelium are the result of a modified cell-nutrition consequent on a morbid condition of blood associated with gout or with one or other of those derangements of the general health to which I just now referred as the usual antecedents of this form of renal degeneration. Gland-cells secreting abnormal products are themselves liable to become abnormal; and, when the process is long continued, the cells are apt to undergo decay and destruction. That appears to me to be the explanation of the disintegration and destruction of the renal epithelium. The wasting and contraction of the tubes, with some real and more apparent thickening of their basement-membrane, are results of the destructive changes in the gland-cells. I cannot explain the replacement of the glandular epithelium by the delicate cells which are found lining some of the tubes, and which, apparently, are intimately associated with the conversion of the tubes into serous cysts.

It seems probable that the copious flow of urine of low specific gravity is a result of the secretion of a watery fluid by the tubes deprived of their normal epithelium, and either denuded or lined by the thin-walled cells before described. This explanation is in accordance with the fact that the urine is usually more copious in the intermediate stages of the disease, when the denuded tubes are most numerous, than in the early stage, when the structural changes in the tubes are less extensive, or in the later stages, when many of the tubes are atrophied and obliterated.

The hypertrophy of the muscular walls of the minute renal arteries is best explained by reference to the analogous phenomena of apnœa. A ligature on the trachea of a dog destroys life in a few minutes; and the immediate cause of death is the arrest of the circulation through the lungs. The chest being opened immediately after death, the left cavities of the heart are found nearly empty; the right cavities, the pulmonary artery to its terminal branches, and the systemic veins, are much distended; the pulmonary capillaries are nearly bloodless; and the lungs consequently collapse to an extreme degree as soon as the chest is opened. Now, what is the explanation of this remarkable and abrupt stoppage of the circulation? The theory is that, when the respiratory changes are suspended by the exclusion of air from the lungs, the minute pulmonary arteries, under the influence of the vaso-motor nerves, so contract as to arrest the flow of blood into the capillaries. It is probable that an impression is conveyed by incident nerves from the pulmonary capillaries to the vaso-motor nerve-centre, whence it is reflected through vaso-motor nerves to the walls of the minute pulmonary arteries, which are thereby excited to contract. Similar phenomena may be observed during a fit of spasmodic asthma. Bronchial spasm narrows the tubes and lessens the supply of air to the lungs; then contraction of the minute pulmonary arteries in a corresponding degree checks the circulation. The skin becomes cold and blue, the pulse small and feeble, and the patient is apparently moribund. When the bronchial spasm relaxes, and the air again gets ready access to the pulmonary vesicles, the arterial contraction ceases, and respiration and circulation together again become free.

You will find the proximate cause of death from apnœa very clearly explained by Sir Thomas Watson in the last edition of his lecture on *different modes of dying*. Now phenomena precisely analogous occur in the kidney and probably in all glandular structures. There is the same intimate relation and interdependence between circulation and secretion as there is between circulation and respiration; in fact, the lung may be looked upon as a gas-secreting gland. When the secreting tissue of the kidney is partially destroyed, the gland is reduced to the condition of a lung receiving only a scanty supply of air, the working power of the gland is lessened, and it requires less blood. The minute renal arteries by their contractile power, under the influence of the vaso-motor nerves, now regulate the blood-supply in accordance with the diminished requirements of the gland. This regulating contraction continues and increases, month after month, year after year; and the physiological result of this persistent over-action of the minute renal arteries is that their muscular walls become hypertrophied. I will show you, hereafter, that a similar hypertrophy of the renal arteries occurs in other forms of chronic Bright's disease, but it is most constant and most conspicuous in the contracted granular kidney which we are now considering. The comparatively small amount of albumen in the urine, and its occasional absence in cases of contracted kidney, may be explained by the fact that, while there is but little compression of the inter-tubular capillaries by swollen tubes, and consequently but little passive engorgement of the Malpighian capillaries, the hypertrophied renal arteries, by their powerful con-traction, prevent a too forcible influx of blood. Hence, too, it happens that hæmorrhage into the tubes, which is so common in acute Bright's disease, rarely occurs in this chronic form of the malady.

The Local and General Symptoms of Contracted Gra-nular Kidney.— There are few diseases equally serious whose progress is so insidious as that of the disease which we are now considering; yet there are few maladies whose presence is indicated by more unequivocal signs, if only they be diligently and intelligently sought for.

One of the earliest symptoms, in the majority of cases, is *increased frequency of micturition,* and especially during the night. The more frequent call to empty the bladder may be a result of a more copious secretion of urine and consequent distension of the bladder, or it may be due to irritation of the bladder by some abnormal quality of the secretion. This symptom is sometimes absent, and it may result from other causes than renal disease. When present, it serves to direct attention to the urinary organs ; and it is a symptom which should never be neglected.

Pain in the back is not a frequent or an important symptom. In many cases it is entirely absent ; and often it is not spoken of until the patient's attention has been directed to the subject. When present, it is more fre- quently muscular than renal—an aching pain in fatigued and feeble lumbar muscles, and often complained of by debilitated patients who have no renal disease. In nume- rous instances, a patient in the advanced stage of incurable degeneration of the kidney has said, ' I cannot understand how my kidneys can be diseased, since I have never had pain in them.'

Dyspepsia is frequently associated with this form of dis- ease, sometimes as a cause, sometimes as a consequence. You may often learn that a patient of strictly temperate habits has for months or years suffered from pain or uneasiness after food, flatulent distension of stomach and bowels, occasional nausea and vomiting, habitual looseness or irregularity of bowels, constipation and diarrhœa alter- nately. With this, there is often turbidity of the urine, which is high-coloured, excessively acid, and deposits urates abundantly. After a time, the urine, which had been scanty, becomes more copious, of pale colour, of low specific gravity, and is found to contain albumen and granular casts. In such a case, probably renal degeneration is a consequence of the long continued elimination of products of faulty digestion through the kidneys. I have seen this sequence of events so frequently, that I have no doubt as to their causative relationship. Dyspeptic symptoms such as I have described, and consequent renal degeneration, are in some cases excited or greatly aggravated by habitual excess of alcohol—less frequently, perhaps, by excessive smoking of tobacco.

In other cases, dyspepsia is a *consequence* of advanced renal degeneration. Urea and other urinary products are vicariously excreted by the mucous membrane of the stomach and bowels, in consequence of the defective action of disorganised kidneys. The gastric secretions are deranged; the digestive functions are disordered; and nausea, vomiting, and diarrhœa, are amongst the results of this secondary renal dyspepsia.

The chronic degeneration of the kidney which we are now considering is often preceded and accompanied by such symptoms as the following : a gradual loss of energy, with emaciation to a variable extent; unusual fatigue after exertion, with a tendency to rheumatic pains and cramp in the feeble and fatigued muscles; defective perspiration, with a dry and harsh state of the skin ; a peculiar pallid or sallow complexion ; and a watery condition of the conjunctiva, or of the connective tissue beneath it. Pallor is not a constant symptom ; there is sometimes a florid complexion even in the advanced stages of this form of degeneration. The tongue is sometimes dry; at other times, moist and pale. There is often thirst, with loss of appetite and some of the dyspeptic symptoms before mentioned. Not unfrequently there is pain or a sense of weight in the head ; sometimes a tendency to drowsiness, and occasional dimness of sight.

Whenever symptoms such as I have described are complained of, the urine should be carefully examined. I need not repeat what I said in the earlier part of this lecture of the indications afforded by the urine from the earliest to the latest stage of this form of renal degeneration.

Dropsy, as I have before told you, is not a prominent symptom in this form of disease. In the majority of cases

F

it is absent throughout the whole progress of the malady. Excluding those cases in which there is the complication of valvular disease of the heart, I found that, of thirty-three fatal cases of contracted kidney, there had been dropsy in only fourteen, the proportion being 42 per cent.; and in most of these fourteen cases the dropsy was only slight and partial, coming on towards the close of the illness. (See a paper on the Forms and Stages of Bright's Disease, ' Med.-Chir. Trans.,' vol. xlii.) The comparative infrequency of dropsy is explained by the free and often copious secretion of urine, which, as a rule, is not highly albuminous. There is not so great a deficiency of the normal blood-constituents in this form of disease as in most acute cases and in other forms of chronic disease. The specific gravity of the blood-serum is less reduced, and the proportion of water to solids is less excessive. An excess of urea, however, is often found in the blood, especially in the later stages, when the secretion of urine becomes scanty.

Hypertrophy of the heart occurs in a large proportion of cases of contracted kidney when the disease has reached an advanced stage. In some cases valvular disease, in others atheromatous and calcareous degeneration of the walls of the large arteries suffices to explain the hypertrophy; but in other cases, as Dr. Bright pointed out more than thirty years ago, there is no such obvious explanation of the hypertrophy, which affects chiefly the left ventricle; and he suggested, as a probable explanation, that ' the altered quality of blood so affects the minute and capillary circulation as to render greater action (of the left ventricle) necessary to force the blood through the distant sub-divisions of the vascular system.' ('Guy's Hospital Reports,' vol. i.) About six years ago, it occurred to me that the hypertrophy of the left ventricle of the heart in cases of contracted kidney might be a result of increased contraction of the small arteries throughout the body, this contraction being excited by the abnormal quality of the blood; and I went on to argue that, if this were so, we should find evidence of the fact in the existence of hypertrophy of the muscular walls of the minute arteries in various tissues. And we have found this hypertrophy not only in the arteries of the kidney, but also in those of the skin, the intestines, the muscles, and the pia mater. It probably exists in the arteries of other tissues which we have not examined.

The probable explanation of the hypertrophied left ventricle in the advanced stage of contracted kidney, then, appears to be this. In consequence of degeneration of the kidney, the blood is morbidly changed. It contains urinary excreta, and it is deficient in some of its own normal constituents. It is, therefore, more or less unsuited to nourish the tissues, and probably more or less noxious to them. The minute arteries throughout the body (of course under the influence of the vaso-motor nerves) resist the passage of this abnormal blood, and in consequence the left ventricle beats with increased force to carry on the circulation. The result of this antagonism of forces is, that the muscular walls of the arteries, and those of the left ventricle of the heart, become simultaneously hypertrophied.

Now I wish to direct your attention to the fact that hypertrophy of the left ventricle, indicated by the apex beating below and external to its normal position, with a strong heaving impulse, and the second sound accentuated over the aortic valves, without signs of valvular disease or senile degeneration of the arterial walls, but with a full resisting radial pulse and high arterial tension, may be taken as evidence that the renal disease is not only chronic, but also in an advanced stage. These physical signs, therefore, will assist you in forming a prognosis.

Both the investing and the lining membrane of the heart are liable to become inflamed, as a result of blood-contamination during the progress of the renal degeneration. This complication will be indicated by the local,. general, and physical signs of pericarditis or endocarditis, or, it may be, of both combined. Other serous membranes sometimes become inflamed—the pleura more frequently than the peritoneum. Œdema of the lungs and bronchitis are frequent complications. Pneumonia is comparatively rare, but it does sometimes occur.

Hæmorrhage.—In the advanced stages of the disease, hæmorrhage from different mucous surfaces is a common and often a troublesome and alarming symptom. Epistaxis is the most common form of hæmorrhage ; but I have seen it occur from the stomach and intestines, from the lungs, the bladder, and from the uterus in the form of menorrhagia. Amenorrhœa is, however, according to my experience, a more frequent result of advanced Bright's disease than menorrhagia.

Cerebral Hæmorrhage.—The most serious and by no means the least frequent form of hæmorrhage is that which takes place within the cranium. In a large proportion, probably half, of the fatal cases of sanguineous apoplexy, the kidneys are found more or less diseased ; and the granular degeneration which we are now discussing is the form of disease which is most frequently associated with cerebral hæmorrhage. The explanation of this common and too often fatal accident is not difficult. The minute cerebral arteries resist the passage of the abnormal blood, while the hypertrophied left ventricle is forcibly driving it onwards. Meanwhile, the walls of some of the intermediate arterie undergo atheromatous degeneration—partly, perhaps, in consequence of the circulation of morbid blood,. partly as a result of the unusual strain and pressure to which they are subjected. At length, in the struggle between the propelling left ventricle and the resisting muscular arterioles, a brittle artery gives way, and a fatal hæmorrhage occurs.

Impairment of vision is one of the most serious results of granular contraction of the kidney. It occurs in two distinct forms : 1. The impairment of vision may be so sudden in its onset, that in a few minutes or hours there is complete blindness, which usually passes away as suddenly as it came. The attacks of sudden and transient blindness may recur again and again. In these cases, ophthalmoscopic examination discovers no structural change within the eye. This form of amblyopia is believed to be of uræmic origin, and is designated uræmic amaurosis. It is usually associated with other symptoms of uræmia, and I shall presently have something more to say of its pathology. 2. In the second form of impaired vision, the dimness of sight comes on more slowly, and is more durable. One eye alone may be affected, but both are often implicated simultaneously or in quick succession. The ophthalmoscope reveals peculiar structural changes in the eye, the result of what is called *retinitis albuminurica*. You will find these appearances fully described and depicted in works on diseases of the eye, amongst which I may especially mention the elaborate and excellent treatise of my colleague Soelberg Wells. Dr. Clifford Allbutt, too, in his able and instructive book on the Ophthalmoscope, has an interesting chapter on the Retinitis associated with Albuminuria. The most characteristic ophthalmoscopic appearances are a broad glistening white mound surrounding the optic disc, the result of sclerosis of the optic nerve fibres and fatty degeneration of the connective tissue elements. The extreme margin of the white mound is broken up into small irregular patches, which assume in the neighbourhood of the yellow spot a peculiar stellate arrangement. The retinal arteries are diminished both in size and number, while the veins are dilated and tortuous. Blood-extravasations, varying in number and in size, sometimes both numerous and large, occur here and there, chiefly in the internal layers of the retina, but sometimes in the external layers, or between the retina and the choroid. The coats of the

blood-vessels are sometimes found in a state of sclerosis or
fatty degeneration. These structural changes appear to be
of an inflammatory and degenerative character. They are
associated more commonly with the contracted kidney
than with other forms of chronic Bright's disease. So
characteristic are the appearances in the retina, and so
insidious is the disease in the kidney, that an ophthalmoscopic examination for determining the cause of dimness
of sight has in many instances led to the discovery of
an unsuspected renal disease. It may be well to mention
here that the two forms of impaired vision which I
have described may occur together or in succession in
the same subject. Uræmic amaurosis may in time be
succeeded by albuminuric retinitis; and the dimness
of vision which results from the latter may be temporarily much increased by uræmic amaurosis. The hæmorrhage into the retina may be explained partly by
the injecting force of the hypertrophied ventricle, partly
by degeneration of the walls of the retinal vessels, and
partly by venous engorgement consequent on pressure upon
the veins by inflammatory exudation.

Cerebral Symptoms. In the advanced stages of contracted kidney, various forms of nervous disorder occur
with so great frequency, that the disease may be said to
have a natural tendency to terminate with symptoms
referable to the brain. These nervous symptoms are very
variable. In some cases, epileptiform convulsions or pro-
found coma may occur suddenly, without premonitory
symptoms. Much more frequently these formidable symptoms are preceded for a variable period by other indications
of brain-disturbance. Amongst the commonest of these
are headache more or less severe and constant, sudden
transient vertigo, equally sudden and transient loss of
sight or hearing, temporary inability to speak, or the
speech for a time is imperfect and stammering; numbness
or neuralgic pains, cramps, chorea-like twitchings, or

transient loss of power, may occur in one arm or leg, or in both the arm and leg on one side; there may be confusion of thought, impairment of memory, and an indescribable nervous dread, with a feeling of utter prostration ; after one or more of these symptoms have continued for a variable period, or recurred more or less frequently, the secretion of urine being usually scanty, and vomiting of frequent occurrence, the patient perhaps becomes drowsy, with more or less delirium ; the tongue is brown and dry ; the breath has a most characteristic sour and fœtid odour ; the drowsiness gradually increases and deepens into coma; the pupils being natural or equally dilated, and the breathing more hurried than in ordinary cases of sanguineous apoplexy ; and so death occurs either with or without convulsions. In some cases, a single attack of violent convulsion is immediately fatal ; in others, the convulsions recur again and again for several hours before the fatal termination. The brain after death is usually found extremely pale and anæmic, with some serous effusion beneath the arachnoid and into the ventricles. These are cases of so-called ' serous apoplexy ;' but the amount of serous effusion is insufficient to compress the brain, and so to explain the symptoms.

Theory of Urœmia.—In attempting to explain these nervous symptoms, I assume it to be indisputable that they are the result of the blood being deteriorated, partly by diminution of its normal constituents, but chiefly by retention and accumulation of urinary excreta. There are two ways in which it is probable that the brain and its functions may be injuriously affected by this blood-deterioration : First, the cerebral tissues, fed with poor and poisoned blood, may have their nutrition impaired, and may in various parts undergo structural changes analogous to those which are often demonstrable in the texture of the retina. Second, it is probable that some of the cerebral symptoms, more especially those which come on and pass away suddenly, are directly due to temporary interruptions or hindrances of the circulation through certain regions of the brain, consequent on excessive contraction of the minute arteries. In my lecture on the Pathology of Epilepsy (published in the ' British Medical Journal,' March 21st, 1868), I adduced many facts and arguments in support of the theory that the immediate cause of an ordinary epileptic convulsion is sudden and extreme anæmia of the brain, the result of excessive contraction of the minute cerebral arteries.

Our increasing experience of the various forms of nervous disorder which may result from so purely mechanical a cause as embolism, in vessels of various sizes and in different regions of the brain, gives additional support and probability to the theory, that many of the cerebral symptoms resulting from uræmia may be explained by a defective blood-supply to certain parts of the brain, consequent on arterial contraction. An arrest of the circulation through a portion of the brain involves immediate suspension of function in that part, with perhaps a disorderly action in subordinate and correlated parts. Thus amongst other symptoms of nervous disorder, maniacal delirium and acute chorea have sometimes been found associated with, and probably have been directly caused by, mechanical plugging of minute cerebral vessels ; the plugging being a result of embolic particles of fibrine detached from the so-called warty vegetations on a damaged mitral or aortic valve. Again, sudden and complete blindness in one eye may result from embolism of the arteria centralis retinæ ; partial and patchy blindness from embolism in one of its branches. It is in a high degree probable that uræmic vertigo, amaurosis, delirium, convulsions, and even coma, may in some cases be explained by partial or general cerebral anæmia, the result of excessive arterial contraction excited by the presence of impure blood, acting through the vaso-motor nerves upon the blood-vessels. I do not ask you to adopt this as a complete and final explanation of the phenomena, but suggest it as a theory to be tested by the results of further observation and research.

Let me add that in some cases, notwithstanding the scantiness and ultimately the almost complete suppression of urine, uræmic symptoms are almost entirely wanting, and consciousness remains until death occurs from exhaustion. In some at least of these cases the uræmic symptoms are probably prevented by the occurrence of incessant vomiting or purging, which, while it rapidly exhausts the patient, favours the escape of noxious impurities from the blood. The cessation of the discharges is sometimes quickly followed by symptoms of uræmia.

Nervous Dyspnœa.—A common and very distressing symptom in the advanced stages of the disease is a peculiar form of dyspnœa. I am not now referring to the persistent dyspnœa which results from the œdema of the lungs, from hydrothorax, or hydropericardium, but to dyspnœa coming on in paroxysms, and especially at night. In some cases the attack resembles asthma, and there are loud sibilant sounds, apparently the result of bronchial spasm ; while in other cases the heart's action is rapid and feeble, and the breathing hurried and laborious, with loud puerile respiration over the lungs. There is evidently no want of moving air in the lungs, and the disturbed circulation and breathing appear to result from some morbid influence of the poisoned blood upon the nervous centres. This distressing form of dyspnœa, which recurs in paroxysms night after night, is, in fact, a form of uræmia.

Disease of the Liver.—In a large proportion of fatal cases of contracted kidney, the liver is found more or less diseased, sometimes enlarged and indurated or fatty, more commonly cirrhosed and contracted. Alcoholic excess may, and often does, excite at the same time cirrhosis of the liver and granular contraction of the kidney. With the cirrhosed liver there is often ascites. When ascites exists without anasarca, or remains after the removal of anasarca, and so forms the prominent dropsical symptom, serious disorganisation of the liver may always be suspected.

Diagnosis.—In addition to what I have said of the symptoms and progress of the disease, I have yet some hints to give you on the subject of diagnosis. The state of uræmic stupor or drowsiness, with a dry tongue and sordes on the teeth, may be mistaken for typhus or enteric fever. The difficulty of diagnosis is increased by the fact, that in some cases of typhus and enteric fever, when there is much cerebral oppression, the urine is often scanty and albuminous, and it sometimes contains granular casts. A close attention to the entire history of the case, and a careful examination of the urine, will seldom leave you in doubt. The specific fever-eruption, when present, is decisive. The thermometer will assist you. The temperature is higher in fever than in uncomplicated uræmic poisoning. Bear in mind that a patient with chronic renal disease may, in addition, have a specific fever—a complication which is usually fatal. With regard to indications afforded by the urine, remember this, that although during the progress of typhus or typhoid fever there may be an acute and transient disintegration of the renal gland-cells, as indicated by the appearance of granular casts, not easily to be distinguished from those which occur in cases of chronic desquamative disease—yet there is this difference, that whereas in the advanced stages of chronic desquamative disease the urine is pale and of low specific gravity, the albuminous urine of fever is usually of deep colour and of rather high specific gravity. It is important to bear in mind that granular casts, with albumen, may appear temporarily in the urine as a result of other blood-poisons than those of typhus and enteric fever. I have seen them in cases of pneumonia, erysipelas, and pyæmia. Once in a case of pyæmia, I found granular and large hyaline cysts exactly like those represented in Fig. 15, but the urine was of deep brown colour and of normal specific gravity; and after death, which resulted from pyæmic abscesses in various parts, the only disease found in the kidney was a recent result of pyæmia. You see, then, that, although the observation of the various forms of tube-casts is of great practical value as an aid to diagnosis and prognosis, yet a too exclusive reliance upon this microscopic evidence may mislead you. When, after a careful inquiry into the history of a case, a doubt exists as to renal disease being recent or of long standing, the evidence of hypertrophy of the left ventricle of the heart without valvular disease, but with a full and firm radial pulse, points to chronic disease in an advanced stage.

I have seen several cases of subacute renal disease occurring in men about middle age as a result of overwork and anxiety, in which it was difficult to decide between acute and chronic disease. I have preserved the urinary sediment from three of these cases, and although the first case occurred nearly fifteen years ago, the tube-casts are as well seen as when the specimen was recent. You may see these specimens under the microscopes on the table, and having carefully inspected them, you may recognise their like when you meet with them in practice. One case was that of a solicitor aged 40, another a merchant aged 56, another a clergyman aged 45. The symptoms and the condition of the urine were alike in all. There was great prostration, vomiting, bleeding at the nose, and in one case from the gums, no dropsy, ultimately a typhoid condition, and unconsciousness shortly before death. The urine was blood-tinged, the specific gravity from 1009 to 1017, moderately albuminous. A rather copious sediment was composed of dark granular and large hyaline casts, with scattered blood discs. Some of the granular casts had a blood tinge, and it is probable that they were in part composed of disintegrated blood. After death, in the only case examined, the kidneys were found somewhat enlarged, soft, and congested. Some tubes were injected with extravasated blood, and others, opaque, with desquamated and disintegrated epithelium. In cases of this kind, although the prognosis is very unfavourable—in fact, all the cases that I have seen have died—yet the disease is not so inevitably fatal as chronic desquamative disease in an advanced stage, and therefore it is important to distinguish between them.

Prognosis.—On the subject of prognosis I have but little to add to what I have already said. Chronic desquamative disease, as a rule, tends gradually to a fatal termination. The rate of progress varies much in different cases and at different periods of the same case. You will remember what I said as to the evidence to be derived from the amount as well as the character of the sediment in the urine. The most trustworthy prognostic indications are to be obtained by comparing the state of the urine with the general symptoms. When with a condition of urine indicating advanced degeneration of the kidney, there is evidence of hypertrophy of the left ventricle, with an unperspiring skin; when with a diminishing secretion of urine, or even without a marked decrease, symptoms of uræmia begin to appear, the disease is generally not far from its fatal termination. You cannot be too cautious in giving a prognosis. The symptoms of chronic renal disease are sometimes much aggravated for a time by some imprudence in diet, by fatigue or anxiety, or exposure to cold. The patient may apparently be on the verge of uræmic coma, or he may have a fit of convulsions, yet, under appropriate treatment, these formidable results of his indiscretion, or his misfortune, may pass away, and, in a few days, he may be apparently no worse than he was before the occurrence of this temporary disturbance. The uræmic symptoms which are not traceable to an obvious external exciting cause are, as a rule, more serious and intractable than those which result from influences capable to some extent of being removed or counteracted.

LECTURE IV.

CHRONIC BRIGHT'S DISEASE WITH A LARGE WHITE KIDNEY.

General History of the Disease: its Causes and Progress—Condition of Urine in different Stages—Various Forms of Tube-Casts, and their Significance—Morbid Anatomy and Pathology of the Kidney in the three Stages of—1. Simple Enlargement—2. Granular Fatty Degeneration— 3. Atrophy, with Coarse Granulations on the Surface—Symptoms—Dropsy —Pulmonary Complications—Inflammation of the Serous Membranes— Endocarditis—Dyspepsia —Vomiting—Diarrhœa — Hypertrophy of the Heart — Cerebral Symptoms—Hæmorrhage, etc.—Defect of Vision— Hæmorrhage from Mucous Membranes—Diagnosis—Prognosis—SIMPLE FAT KIDNEY, OR GENERAL FATTY INFILTRATION OF THE KIDNEY—History —Microscopic Characters of the Kidney—Pathology and Clinical History —Points of Difference between it and the Granular Fat Kidney.

I NOW propose to give you the pathological and clinical history of those cases of chronic Bright's disease in which the kidneys are found, after death, always more or less anæmic and pale, usually enlarged, soft in consistence, and smooth on the surface, but sometimes contracted, indurated, and coarsely granular.

A large white smooth kidney is often a sequel of an acute inflammatory attack. Acute Bright's disease, the result, it may be, of scarlet fever or of exposure to cold, is imperfectly recovered from. The dropsy passes away ; the patient regains his strength and his colour ; the urine is normal in quantity, appearance, and specific gravity ; but it continues to be more or less albuminous. The albuminuria which thus continues after acute Bright's disease may remain for many months, and even for a number of years, before the appearance of symptoms con-

sequent on chronic and incurable degeneration of the
kidney. At length there is, perhaps, a return of dropsy,
or the patient is cut off by some of the results of uræmia,
either of an inflammatory or neurotic character. After
death, the kidneys may present one of three distinct appearances: 1. They may be large, almost uniformly white,
and smooth on the surface. This appearance is admirably
represented in Dr. Bright's fourth Plate, Figs. 1 and 2.
I call this simply a 'large white kidney.' 2. The kidney
may present the same general appearance, but with this
addition, that the cortical surface and the surface of a
section of the cortex are, to use the words of Dr. Bright,
'interspersed with numerous small yellowish opaque specks.'
This appearance is represented in Fig. 3 of Dr. Bright's
third Plate. I shall presently show you that these yellow
opaque specks are spots of fatty degeneration ; and I call
this a ' granular fat kidney,' or a ' large white kidney with
fatty granulations.' 3. The cortical portion of the kidney
may be found more or less atrophied, with an uneven
granular surface ; the yellow specks of fatty degeneration
being in some cases still visible on the surface and on
section.

The various appearances which I have described are results of successive stages of the same pathological process ;
in the majority of cases, but not always, following upon an
acute onset. A kidney, which has contracted and become
granular, after having been enlarged, differs from the
contracted granular kidney which I described in my last
lecture, in being of paler colour, of firmer texture, and
more coarsely granular on the surface. The microscopic
appearances of the kidney, the history, and the symptoms,
also differ greatly, as we shall presently see.

There is a class of cases in which, with a clinical history
different from that of the cases to which I have just now
referred, the kidneys are found, after death, pale and wax-
like in colour and consistency, usually enlarged and smooth,

but sometimes contracted and granular. This is the 'lardaceous degeneration' of the kidney. The subjects of this form of disease have usually been strumous or otherwise cachectic before the onset of the renal disease, which, in the great majority of cases, begins as an insidious chronic malady. A similar degeneration of the liver and spleen is often associated with that of the kidney. In my next lecture, I shall discuss the clinical and pathological history of this lardaceous degeneration of the kidney.

I now proceed to give you a more detailed account of those cases which are associated with a large white kidney, with or without the fatty granulations, with or without subsequent contraction and coarse granulations on the surface.

In a large proportion of cases, the commencement of the disease dates from an attack of acute Bright's disease resulting from exposure to cold, scarlet fever, diphtheria, typhus or enteric fever, or other zymotic disease. In several instances, I have traced the disease back to an attack of tropical malarious fever; I have seen it as a sequela of ague in this country; and I have known it follow upon an attack of dysentery. The acute disease may or may not have been associated with dropsy. The dropsy, if present, usually passes away, and for a time the only evidence of incomplete recovery is to be found in the condition of the urine. In some few cases, the disease comes on as an insidious chronic malady, and it is impossible to determine either the date of its commencement or its probable cause.

It is not unfrequently a result of, or, to say the least, it is often associated with, an excessive consumption of food and of alcoholic stimulants, with consequent dyspepsia. In cases thus originating, the approach of the disease is gradual, insidious, and often unsuspected until it has reached an advanced stage. The disease occurs at all ages, from infancy to extreme old age. I have seen it fatal at the age of seventy-five. Beyond childhood, it is more com-

mon in males than in females—probably because males
are more exposed to cold, and more intemperate, than the
other sex. This form of disease has, as a rule, a more protracted course than any other form of Bright's disease ;
and for a period of years it may be unattended by any
obvious symptoms apart from the indications afforded by
the urine. A patient recovers from the dropsy and other
symptoms of acute Bright's disease ; he feels and declares
himself to be quite well ; but the urine shows that recovery
has not been complete. The urine may for a long time
be normal in quantity, colour, and specific gravity ; but it
contains albumen, varying in amount from a mere opales-
cence, with heat and acid, to a dense and copious precipitate ; the albumen being usually more copious after food
and exercise. The urine, placed in a conical glass, may
remain quite clear, and show no appearance of tube-casts
or renal epithelium ; or it may either occasionally or constantly deposit a light cloud, in which are found small
hyaline casts, in some of which a few oil-globules may
perhaps be seen, while others contain an epithelial cell or
two, or some fragments of cells. This condition of urine
may continue with little or no change for many months,
and even for several years, before there is any indication
that the general health is suffering from the state of the
kidneys. The scarlet fever poison and its products, which
originally excited the acute renal disease, have passed away,
a certain amount of injury having been inflicted upon the
kidney, from which it has not recovered ; but there is, up
to a certain period, no progressive disease of the gland.
The convoluted tubes have been left enlarged ; the cortex
probably has a somewhat anæmic and mottled appearance,
resulting from compression of the intertubular capillary
network by swollen tubes. This compression of the intertubular capillaries to some extent impedes the circulation ;
there is consequent engorgement of the Malpighian capillaries, and hence a transudation of serum into the tubes,

which, mingling with the urine, renders it albuminous. This explanation of the albuminuria will readily be understood by a reference to Fig. 1 in my first lecture. Its mode of production is analogous to that of the albuminuria and the hæmaturia which Dr. George Robinson first, and Frerichs and others since, have produced in rabbits by putting a ligature on the renal vein. In addition to this mechanical hindrance to the passage of blood, it is not improbable that, as a result of acute Bright's disease, the Malpighian capillaries may undergo some physical change, which for a time, or even permanently, favours the transudation of serum through their walls. After a variable period of months or even years, the albuminuria may at length cease, and the cure is complete ; but persistent albuminuria, or the state of kidney which gives rise to it, involves, sooner or later, serious structural changes in the kidney, and, as a result of these changes, some of the secondary results of renal degeneration. The state of things which results from persistent albuminuria following the incomplete cure of acute Bright's disease, is analogous to that which occurs when acute endocarditis has passed away, but has left a thickened and defective valve. For a time, hypertrophy of the heart's walls compensates for the imperfect valve, and the circulation appears to be unimpeded ; but there is a limit to this conservative process. At length, the muscular tissue of the heart ceases to respond to the increasing demands which are made upon it, and it undergoes degenerative changes, the circulation flags, and the serious trouble begins. In like manner, for an indefinite time, the urine, although more or less albuminous, is freely secreted, and contains its due proportion of solids to liquid, there being no indication or symptom of defective secretion. Sooner or later, however, the urine loses its natural sherry-colour, and gradually becomes paler. The secretion is less copious. The specific gravity rises. With a scanty secretion, it may be as high as 1020, or

G

even 1030 ; but, with a more copious secretion, it may be
as low as 1010. By this time, probably, the sediment in
the urine will have become more copious ; and it may contain small hyaline and oily casts and cells in considerable
numbers. These appearances indicate with absolute cer-
tainty that, in some of the uriniferous tubes, the gland-
cells are undergoing fatty transformation ; and the kidney
will present, after death, those small yellow spots of fatty
degeneration to which I just now referred.

The kidney has now passed from the first stage of a
simple large white kidney to the second stage—namely,
that of a granular fat kidney ; and the disease is usually
fatal in this stage ; but it may pass on to the third stage
of contraction and atrophy, with coarse granulations on
the surface, the small yellow fat granulations being still
visible. This stage is indicated by the appearance of large
granular and large hyaline casts in the urine. The large
size of these casts shows that they were moulded in tubes
which have been deprived of their lining of gland-cells.
The destruction of the gland-cells is followed by atrophy
of the gland ; and the amount of urinary sediment, containing the large granular and large hyaline casts, is an
index of the rate at which the disease is making progress.
The granular casts in part consist of disintegrated epithelium—in part of the disintegrated fibrinous material of
which the hyaline casts are composed.

I have in many instances traced the transition from
acute to chronic disease ; and in chronic cases I have
traced, by the microscopic appearances and other physical
changes in the urine, the successive stages of a large white
smooth kidney, a granular fat kidney, and, lastly, a contracted and coarsely granular kidney. This sequence of
events occurred in a case the later stages of which are
illustrated by Figs. 25, 26, and 27. The disease began in
a man aged 23, as an acute attack, in October 1846. It
passed into a chronic stage, with oily casts and cells, which

continued for a period of nine years. Then the oily casts and cells were mixed with, and afterwards replaced by,

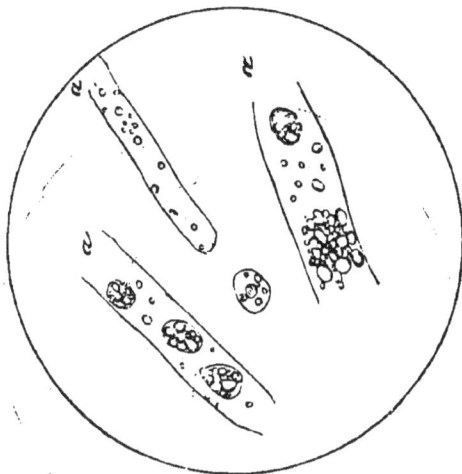

Tube-casts at three successive periods of the same case.
Fig. 25.—Period of Fatty Enlargement of the Kidney.

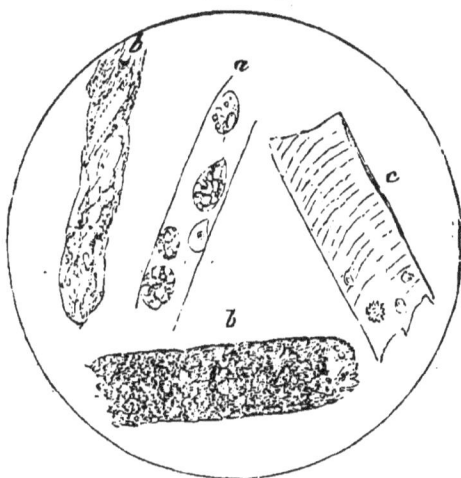

Fig. 26.—Commencing Atrophy and Contraction.

large granular and large hyaline casts. Death occurred from uræmia in October 1856, ten years after the onset of

G 2

the disease. The kidneys were pale, had many yellow fat granulations in the cortex, and were much atrophied, their combined weight being only 7¾ ounces. Their appearance is represented by a chromo-lithograph illustrating a paper of mine in the forty-second volume of the 'Medico-Chirur-

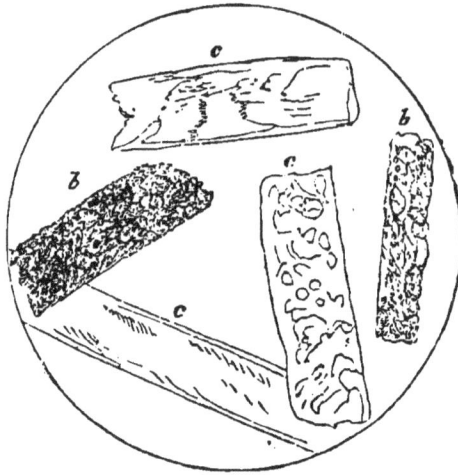

Fig. 27.—Advanced Atrophy and Contraction. *a a a a.* Oily Casts and Cells. *b b b b.* Granular Casts. *c c c c.* Large Hyaline Casts.— × 200.

gical Transactions.' The microscopic specimens from which Figs. 25, 26, and 27 are taken, retain their characteristic appearances after an interval of seventeen years; and they are placed for your inspection beneath the microscopes on the table.

Morbid Anatomy and Pathology of the Kidney.— When death has occurred before oily casts and cells have appeared in the urine, the kidneys will be found in the first stage of degeneration—that is, large, white, and smooth on the surface. The weight of each kidney may be from seven to ten ounces. The cortical portion is increased in thickness, and appears more or less anæmic ; while the medullary cones are pink and vascular. The lobular markings on the surface, which are in fact the radicles of the renal vein, are more or less obliterated, so that perhaps there remain only a few isolated vascular patches, as represented in Dr. Bright's fourth plate. Rarely some hæmorrhagic spots are scattered over the surface or through the substance of the cortex. The capsule readily peels off, and leaves a smooth surface. On a microscopic examination, the greater number of the convoluted tubes present no other change than that of being larger and more opaque than usual. The epithelium is unusually granular and opaque, and apparently contains a more than ordinary amount of solid matter; while the central axis of the tube is lighter than the margins, and free from deposit or accumulation. The appearance of the tubes is precisely the same as that which I described in cases of acute Bright's disease unassociated with epithelial desquamation (see *ante*, Lecture II. Fig. 12). In some tubes, the central canal contains the fibrinous material which appears in the urine in the form of small hyaline casts. The hæmorrhagic spots, when present, are seen to be convoluted tubes filled with blood from ruptured Malpighian capillaries (see *ante*, Fig. 11). The Malpighian capillaries usually have their walls more or less thickened and opaque, and they often have a wax-like appearance. The muscular walls of the minute arteries are sometimes hypertrophied ; but arterial hypertrophy is not constant in this stage of the disease. As a rule, the hypertrophy of the arterial walls bears an inverse relation to the general enlargement of the kidney. I shall presently suggest an explanation of this fact. The intertubular capillaries and veins present no structural change, but they are much compressed by the enlarged tubes ; and this explains the disappearance of the lobular markings from the surface of the kidney. The enlargement of the cortical portion of the kidney is mainly a result of a kind of hypertrophy of the gland. Many of the tubes are certainly enlarged, and their epithelium is unusually opaque and bulky. The transverse sections of many tubes have twice the normal diameter, and these large tubes must

contain an increased number of secreting cells. This increase in the diameter of the secreting tubes is analogous
to the increased thickness of fibre in a growing and hypertrophied muscle. When one kidney has been destroyed
by disease or accident, as by the impaction of a calculus in
the ureter, the other does double work, and, in so doing,
doubles its size, without undergoing structural change;
the gland is simply hypertrophied. The pathological enlargement of the kidney is like this, but with a difference.
When acute Bright's disease leaves such an amount of
swelling of tubes and consequent impediment of the circulation as interferes with the prompt and complete excretion
of urine, there will continually be some excess of retained
urinary products in the blood; and this accumulation of
urinary excreta will act as a stimulus to increased growth
and development of glandular tissue. For the tissues in
question may be said to feed upon those materials for
which they have a special affinity; and the growth of a
gland is in proportion to the amount of the materials for
its proper secretion with which it is supplied by the blood.

The hypertrophy goes on up to a certain point, and then
a process of atrophy begins. The commencement of this
is indicated during life by the appearance of oily casts and
cells in the urine; and then, after death, the kidneys are
found in the second stage—namely, that of fatty granulation, as before described, and as represented in Fig. 3 of
Dr. Bright's third Plate.

These yellow specks are usually numerous in the cortex,
but are never seen in the cones. Place a section of a
yellow spot under the microscope, and you see that, as a
hæmorrhagic spot is a convoluted tube filled with extravasated blood, this yellow speck is a tube filled with oil,
mostly within cells, but partly loose. It is evident that,
in this stage, the epithelium in certain sets of tubes has
undergone fatty degeneration (see Fig. 28). The fatty
nature of the material is proved by its solubility in ether,

and by the smaller particles fusing and forming larger globules when gently warmed. The probable explanation of these spots of degeneration is, that the swollen tubes so compress the intertubular capillaries, and thus impede the circulation, that the nutrition of the tubes is impaired, and their cells undergo fatty degeneration. The phenomena are analogous to the softening, with fatty transformation of the brain-tissue, as a result of embolism in a

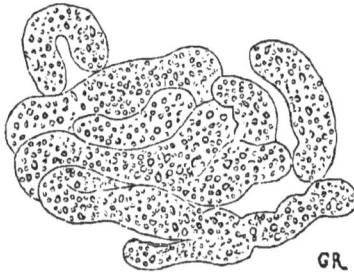

Fig. 28.—A Yellow Speck or Granulation magnified 100 diameters, and thus shown to be a Convoluted Tube, with its contents in a state of Fatty Degeneration.—× 100.

cerebral artery. They also resemble the circumscribed patches of fatty degeneration of the muscular walls of the heart consequent on obstruction of branches of the coronary artery distributed to the diseased parts. (See, upon this point, Dr. Quain's classical paper on Fatty Degeneration of the Heart, ' Med.-Chir. Trans.,' vol. xxxiii. p. 147).

But the degenerative changes may proceed further, and lead to a rapid disintegration of the gland-cells, and their replacement by unorganised fibrine. This change is indicated during life by the appearance of large-sized granular and hyaline casts in the urine (see Fig. 27), and after death by some of the tubes being filled with the same materials. These destructive changes in the gland-cells explain the atrophy and the granular contraction of the kidney in the third stage of the disease. I may mention here that a large white kidney, the result of acute Bright's

disease, sometimes passes on into the stage of atrophy with
a coarsely granular surface, without going through the intermediate stage of fatty degeneration. This I know to
be a fact, from close observation of the urine during life,
and a comparison of its microscopic characters with the
appearances in the kidney after death. In such cases you
may find here and there microscopic evidence of fatty degeneration within a tube; but the change is so slight as
to be invisible by the unaided eye. No fatty granulations
appear upon the surface or on a section of the gland. In
the more advanced stages of this chronic disease, more
especially in the stage of atrophy, the muscular walls of
the arteries are almost always more or less hypertrophied.
The explanation of the fact which I just now mentioned—
that, as a rule, there is an inverse relation between enlargement of the kidney and hypertrophy of the arterial walls—
is, that up to a certain point the increased growth of the
glandular tissues obviates the necessity for that stop-cock
action of the minute arteries which occurs when the glandular tissue is wasting, and which results in hypertrophy of
the arterial walls, as I explained to you in my last lecture.
The rule is, that the arterial hypertrophy commences when
the glandular hypertrophy ceases and is succeeded by
atrophic changes in the gland-cells and tubes. The Malpighian capillaries are usually more thickened, glistening,
and wax-like than in the earlier stages of the disease.
This apparently is a result of the continued transudation
of albuminous and fibrinous materials through their walls,
and their consequent infiltration with these products.

We occasionally find that the walls of some of the renal
arteries have undergone a peculiar change which gives
them a homogeneous wax-like appearance. I shall hereafter describe and explain this form of degeneration in
connection with lardaceous disease of the kidney. In the
stage of atrophy and contraction, some convoluted tubes
may be seen denuded, some being contracted and others

dilated. These appearances are the same as are found in the small red kidney. In the advanced stages of the disease too, the basement-membrane of some of the tubes and the capsules of the Malpighian bodies are somewhat thickened, but less decidedly and constantly than in the contracted granular kidney.

SYMPTOMS.—Amongst the symptoms of chronic Bright's disease with large white kidneys, *dropsy* is one of the most frequent and prominent. The cases in which dropsy is absent throughout the whole progress of the disease, form a very small minority. Some years since I found, from an analysis of twenty-six fatal cases, that there had been dropsy in twenty-four; the proportion being, in round numbers, 92 per cent. And in the majority of cases the dropsy was great, general, and of long duration. (See my paper on the Forms and Stages of Bright's Disease, 'Med.-Chir. Trans.,' vol. xlii.) When chronic disease has supervened upon an acute attack, the dropsy which attended the onset of the malady may continue throughout, or it may pass away and return after a variable interval of months or years ; or, if the urine become copious, as it sometimes does in the second and third stages, the dropsy may never return. Acute renal dropsy usually becomes general within a short period ; but the dropsy which accompanies this chronic form of disease shows itself first in the face and feet, and gradually extends to other parts, including, in the worst cases, the serous membranes of the chest and abdomen. The pale, pasty, and puffed appearance of the face is very characteristic of this form of renal disease. As a rule, the more scanty the secretion of urine, and the larger the proportion of albumen which it contains, the greater is the tendency to dropsy.

When dropsical swelling of the legs becomes excessive, so as to cause great tension of the integuments, the circulation through the skin and the subcutaneous tissues is seriously impeded ; and this may result in erysipelatous

inflammation, sloughing, and gangrene. In one case, I
saw gangrene and sloughing of the skin over the back of
both legs, excited by the pressure of the heavy dropsical
limbs upon the bed.

Congestion and œdema of the lungs are often associated
with, and may be said to form a part of, the dropsical
symptoms. The lungs are not rarely the seat of inflammatory mischief. *Bronchitis* and *pneumonia* are amongst
the more frequent and serious complications. *Submucous
œdema of the larynx* occasionally occurs, and renders the
voice husky ; more rarely it causes stridulous breathing
and dyspnœa. *Inflammation of the serous membranes*
is one of the less frequent complications. According to,
my experience, the pleura is more frequently inflamed
than the pericardium; the peritoneum less frequently
than either. *Endocarditis* is sometimes set up, and may
result in a chronic valvular disease. *Derangements of
the stomach and bowels* are of common occurrence in
all stages of the disease, but more especially in the advanced stages, and when, from any cause, the secretion
of urine becomes scanty. There is loss of appetite ;
dyspepsia, with flatulent distension after food ; nausea ;
water-brash ; and vomiting, especially in the morning,
when it is often excited by the attempt to clean the teeth.
When the urine is scanty, the vomited matters often have
a fœtid ammoniacal odour. In the advanced stages, the
vomiting may be almost incessant and quite irrepressible.
Diarrhœa, too, is not an unfrequent symptom. It may be
excited by ill-digested food, or by the vicarious excretion
of urinary products ; not unfrequently, perhaps, by both
these influences combined. In the advanced stages, the
skin is usually unperspiring, dry, and harsh. *Hypertrophy of the heart* is less frequently associated with this
form of disease than with the chronic desquamative disease.
The more advanced the stage of disease, the more frequent

is the occurrence of hypertrophy. Dr. Grainger Stewart found this condition in only 12 per cent. of the cases fatal in the first stage, in 38 per cent. of those fatal in the second stage, and in 100 per cent. of those fatal in the third stage—that is, the stage of contraction ('A Practical Treatise on Bright's Diseases,' 2nd edition, p. 90). The probable explanation of the comparative infrequency of hypertrophy of the heart in the earlier stages is, that the watery condition of blood which results from a defective secretion of urine excites less resistance in the minute systemic arteries than the more concentrated blood-poison which results from the retention of urinary solids, while the water freely filters off through the small red kidney. For the same reason, too, *cerebral symptoms of uræmic origin* and *cerebral hæmorrhage* are less frequently associated with the large white kidney than with the contracted granular kidney. It appears that, while dropsy results from hydræmia, arterial resistance, hypertrophy of the heart, toxæmic nervous symptoms, and cerebral hæmorrhage, are more direct results of uræmia.

The two forms of *defect of vision* which I described as of frequent occurrence in connection with the small red kidney, are less frequently associated with the disease which we are now discussing. The difference is one of degree. I have seen several cases of uræmic amaurosis and albuminuric retinitis with an unquestionable history, though not with *post-mortem* evidence of an enlarged, white, and fat kidney.

Although cerebral hæmorrhage is a less frequent result of this form of disease than of the chronic desquamative disease, *hæmorrhage from mucous membranes*, and especially from that of the nose, is, in the advanced stages, a frequent and often a formidable symptom. These hæmorrhages are probably in part explained by the blood-deterioration, more especially the deficiency of albumen and hæmoglobin, and in part by the malnutrition and consequent brittleness of the walls of the vessels in the advanced stages of the disease.

Diagnosis.—On the subject of diagnosis, I need not add much to what I have already said. A careful consideration of the general history and the symptoms, together with the characters of the urine, will rarely leave you in doubt as to the form and stage of the disease. Renal disease dating from an attack of acute general dropsy, followed by persistent albuminuria, can rarely be other than the particular form of disease which we are now discussing. The stage of the disease is to be determined chiefly by the character of the urine. In the first stage, the urine is usually normal in quantity, in colour, transparency, and specific gravity. There is either no deposit, or a cloudy sediment containing some small hyaline casts. As the disease advances, the urine gradually loses its sherry tint, and becomes lighter coloured. The amount secreted and the specific gravity usually bear an inverse relation to each other. In the second stage, the small hyaline and oily casts are found; and in the third stage the oily casts are mingled with or replaced by the large granular and large hyaline casts. (Figs. 25, 26, and 27.)

Prognosis.—Your judgment as to the probable result of the malady will obviously be greatly influenced by the opinion which you may form as to the stage of the disease and the rate at which it is making progress. In my lecture on acute Bright's disease, I told you that I have seen cases of complete recovery after oily casts and cells had appeared in the urine continuously for many weeks, and after albuminuria had existed for one, two, and even three years.

The most favourable condition of urine is that in which it retains its normal colour, deposits no sediment on standing, and contains but little albumen. There can be no question that recovery may take place after the disease has passed into the second stage—namely, that of fatty degeneration. You are not, therefore, to despair of a patient whose urine contains oily casts even in large numbers. As a general rule, the longer the continuance of albuminuria in spite of careful treatment, and the greater the amount of albumen, the more unfavourable is the prognosis. In estimating the amount of albumen, never omit to compare the urine after food and exercise with that passed after rest and fasting. When the urine is very pale, and of low specific gravity, yet highly albuminous, when it deposits a copious and dense sediment composed in great part of large granular and large hyaline casts, it will be evident that the disease is in the third stage, and that the kidney is contracting. The number of the large-sized casts may be taken as an index of the rate at which the degenerative and atrophic changes are progressing. The duration of the disease, in cases which ultimately prove fatal, varies extremely. I have notes of the case of a child aged 7, who had acute renal disease with dropsy after scarlet fever ; and, the malady having terminated fatally within five weeks from its onset, the kidneys were already in the second stage, the pale and enlarged cortex being scattered over with the characteristic yellow spots of fatty degeneration. In this case, the disease ran an unusually rapid course. I have seen a considerable number of cases in which the symptoms have continued for from five to ten years. My experience does not accord with the statements of some recent writers, who affirm that this disease is of shorter duration than that which results in the red granular kidney. I have before referred to one case which extended over a period of ten years ; but the most prolonged case that I have seen or heard of was that of a medical practitioner whose history is partly given in my book on the ' Kidney' (p. 374). He had dropsy after scarlet fever in 1836, when he was about seventeen years of age. He recovered from the dropsy, and thought no more

of his malady until five years aftewards, when his urine
was accidentally discovered to be albuminous by a fellow
medical student. It had probably been so since the attack
of dropsy; and it certainly remained albuminous from
that time until his death, which resulted from dropsy in
May 1866. He was then in his forty-seventh year; and,
if we assume, as we safely may, that the urine had not
ceased to be albuminous between the attack of dropsy and
the accidental discovery of the albumen five years later,
this gentleman had albuminuria for thirty years before his
death; yet the greater part of that time he was a hard-
working general practitioner, and, to all outward appearance, in good health. During the last year or two of his
life, his urine was saccharine as well as albuminous. When
I first saw him in 1851, his urine was of normal colour and
specific gravity, but albuminous; it deposited no sediment,
and contained no tube-casts. I believe that for many years
it retained the same characters, but I have no note of any
subsequent microscopical examination. This case teaches
two practical lessons. The first is, not to take for granted
that a patient who has recovered from acute renal dropsy
is well until his urine has lost all trace of albumen; and
the second is, not to assume that persistent albuminuria
of necessity involves early death or the speedy occurrence of
formidable symptoms.

THE SIMPLE FAT KIDNEY OR GENERAL FATTY INFILTRA-
TION OF THE KIDNEY.—Before passing on to the subject of
lardaceous degeneration of the kidney, I wish to direct your
attention for a moment to a condition of the kidney which
may be designated 'the simple fat kidney' or 'general
fatty infiltration of the kidney.'

There is a form of fat kidney very different from that
which I have described as the granular fat kidney. It
consists in an uniform infiltration of the epithelium of the
convoluted tubes with oil. This state of kidney is analogous
to the fatty infiltration of the cells of the liver which occurs

often in cases of phthisis and other wasting diseases. This form of fat kidney I first discovered and described in the year 1846 ('Med.-Chir. Trans.' vol. xxix). The fatty infiltration of the renal epithelium may be found in various grades. When the fat is very abundant, the kidney is increased in size and weight. The colour of the cortex is either uniformly pale, or more frequently mottled by a blending of pale anæmic with red vascular patches. Occasionally, hæmorrhagic spots are scattered through the cortical substance. The medullary cones retain their normal colour and vascularity. The consistence of the kidney is usually softer than natural, and frequently the gland has an œdematous feel and appearance. On a microscopic examination, the convoluted tubes are found to be universally distended with oil which has accumulated in their epithelial cells. There is a uniform oily infiltration of the renal gland cells. (See Fig. 29.) This condition of the kidney is found

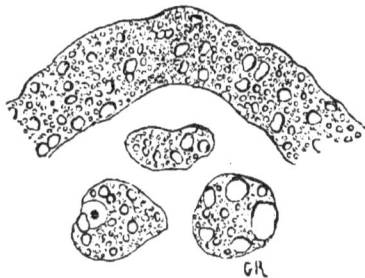

Fig. 29.—A portion of a Convoluted Tube distended with Oil, from a ' simple fat Kidney.' Three detached Epithelial Cells; two are filled and distended with oil, the third contains oil in less quantity, and the Cell-nucleus is visible.—× 200.

not unfrequently associated with a similar condition of liver in persons who have an excess of adipose tissue beneath the skin, in the abdomen and about the heart. It is commonly found after death from diabetes and from some other chronic diseases which are attended with great emaciation, such as cancer, phthisis, and dysentery. It is probable that the immediate cause of this fatty infiltration of the gland-cells is an excess of fatty matter in the blood. In the case of very fat persons, who are usually large con-sumers of fat-making adipose and amyloid food, the materials whence the fat is derived are introduced directly into the blood through the stomach. On the other hand, in cases of wasting disease, it is probable that the fat absorbed from the adipose tissues enters the circulation and infiltrates the gland-cells, more commonly those of the liver, less frequently those of the kidney.

Fat kidneys are common in the domestic dog and cat, probably because these animals lead indolent lives, and consume large quantities of food rich in hydrocarbon. These animals are the counterparts of the human animal when, from eating and drinking to excess, he grows fat and gets fat liver, heart, and kidney. The unhappy Strasbourg geese afford an illustration of fatty infiltration of the liver resulting from a wasting disease. In order to obtain fat livers for patties, the animals are well fed and fattened; then they are confined in heated cages without food and water. They become feverish, and rapidly waste, while their livers grow large. It seems probable, as Baron Larrey long ago suggested, that the oil absorbed from the adipose tissue enters the circulation and infiltrates the cells of the liver, and probably in a less degree those of the kidney also.

It is a remarkable fact, that the liver and kidneys have been found in a state of extreme fatty infiltration in cases of poisoning by phosphorus, death occurring within a week after the poison was taken. Two cases of this kind are recorded in the fiftieth volume of the 'Medico-Chirurgical Transactions'—one by Dr. Habershon, the other by the late Dr. Hillier. Fatty infiltration of the liver and kidney appears, as a rule, to have but little influence on the functions of these glands. We sometimes, however, find jaundice and ascites associated with fat liver, and with no other structural change to explain the symptoms; and I have notes of several cases in which albuminuria and the usual symptoms of chronic Bright's disease have occurred in connection with simple fatty infiltration of the con- voluted tubes of the kidney. One such case I published in my book on 'Diseases of the Kidney' (case of Ann White, p. 414). In that case, the disease followed a second attack of scarlet fever with anasarca. There was dropsy, and the urine contained much albumen and oily cells. She died with uræmic convulsions and coma. The convoluted tubes were greatly and almost uniformly gorged with oil ; and Dr. Beale, analysing the cortex of the kidney, found that more than one-fourth of the solid matter was fat. It is reasonable to suppose that so large an accumulation of oil within the gland-cells must impair their secreting power, and also impede the circulation through the inter- tubular capillaries, which are compressed by the distended and swollen tubes.

I will now briefly recapitulate the chief points of dis- tinction between the 'granular fat kidney' and what I have here called the 'simple fat kidney,' which is perhaps a better term than 'the mottled fat kidney,' which I formerly employed, but which, I am told, has often been misunderstood. In the granular fat kidney, there are disseminated spots of fatty degeneration in the cortex ; and these are secondary results of previous structural changes in the gland. In the simple fat kidney, on the contrary, there is a general fatty infiltration of the gland-cells in the convoluted tubes of the cortex ; and this is a primary change. The granular fat kidney is always associated with albuminuria, and often with other signs of serious disturbance of function. The fatty infiltra- tion, although it is sometimes associated with albuminuria and other symptoms of renal disease, is, in the majority of cases, unattended by obvious signs of functional derange- ment. This condition of kidney, therefore, while it has great interest for the pathologist, has much less clinical ·

importance than the 'granular fat kidney;' but I cannot assent to the statement that, because overfed animals leading unnaturally indolent lives have an excess of oil in their kidneys, this condition is normal or innocuous. By parity of reasoning, it might be maintained that an excessive growth of fat about the heart is a harmless addition of hydrocarbon to the weight of the body, because that state of heart often coexists for a time with apparently good health and great bodily activity.

LECTURE V.

CHRONIC BRIGHT'S DISEASE, WITH A LARDACEOUS OR WAXY KIDNEY.

General History—Virchow's erroneous Theories—Clinical History and Symptoms—Minute Anatomy and Pathology of the Kidney—Its Relation to Continued and Profuse Suppuration—Hypertrophy of the Heart—Diagnosis—Prognosis—Hæmaturia in Chronic Bright's Disease.

THERE are cases of chronic Bright's disease associated with kidneys which are usually enlarged, anæmic, pale, and waxlike; thus resembling in some respects the cases which I described to you in my last lecture. But the form of disease to which I now invite your attention has for the most part a peculiar and distinctive clinical history; and the anatomical condition of the kidney is in some respects different from that of the ordinary large white kidney.

General History.—The subjects of this form of disease have usually been strumous or otherwise cachectic before the onset of the renal degeneration. In some cases, there has been strumous disease of a joint or of one or more bones, with long continued suppuration; in others there are symptoms of phthisis. In a large proportion of cases, there is a history of constitutional syphilis, with resulting cachexia. In some cases cancer, in others dysentery, in others habitual intemperance, in others, again, long continued albuminuria following upon acute Bright's disease, has led to the cachexia out of which this form of renal disease has sprung. One of the earliest symptoms of the disease is a copious flow of urine, at first perhaps not albuminous, but subsequently more or less impregnated

H 2

with albumen.   Another common symptom is profuse and
obstinate diarrhœa.   Dropsy, more or less general, usually
occurs ; but it is not so constant or so prominent a symptom
as in the class of cases which I described in my last lecture
as sequelæ of acute Bright's disease.   When at length the
patient dies, sometimes from uræmia, but more frequently
from exhaustion, the kidneys are found in a state which
has been called ' lardaceous ' or ' waxy degeneration.'   The
gland is usually enlarged, sometimes very much so.   In
one of my own cases, the two kidneys weighed twenty-eight
ounces.   Dr. Dickinson in one case found their combined
weight thirty-three ounces.   The surface of the kidney is
smooth and pale ; the texture of the anæmic and thickened
cortex is firm, and has the semi-translucent appearance of
white beeswax ; while the cones retain their normal colour,
vascularity, and size.   The cut surface presents numerous
glistening points, due to the altered Malpighian capillaries.
In some cases, minute yellow fat-granulations are scattered
through the cortex.   This is the large, smooth, lardaceous
kidney ; and one of its most remarkable and distinctive
features is, that in the majority of cases it is associated
with an analogous condition of the liver or spleen, or of both.
In a certain proportion of cases, a stage of atrophy follows
upon that of enlargement ; the cortical substance wastes,
and coarse granulations appear on the surface.   This is the
' contracted ' or ' granular lardaceous kidney.'   Such a
kidney is represented in Figs. 1 and 2 of Dr. Bright's third
Plate.   It is remarkable that Dr. Bright's plates contain
no illustration of the small red granular kidney, whose
history I gave you in my third lecture.

     Before entering upon a minute description of the kidneys,
it will be well to give a brief history of the disease and of
the speculative doctrines to which it has given rise.   Virchow
is the author of two theories regarding this disease.   The
one theory is, that the blood-vessels are the primary seat of
the degenerative changes; and the other is, that the morbid

deposit is of the nature of starch or vegetable cellulose; and he therefore calls the disease ' amyloid degeneration.' I shall presently show you that the first theory is erroneous ; and the second is now universally admitted to be so. The term ' amyloid' was suggested by a supposed chemical resemblance between the morbid deposit and vegetable cellulose or starch, as shown by the staining with a solution of iodine; but careful analyses by various competent chemists have shown that the material has essentially the same composition as the protein compounds, and that it is of an albuminous or fibrinous nature.

It is very desirable that the term amyloid, which is based upon an erroneous chemical theory, should be discontinued ; and that the term ' lardaceous,' recommended by a Committee of the Pathological Society (*Path. Trans.* vol. xxii. p. 2), should be adopted. The term lardaceous means no more than that the disease has the appearance of bacon fat (*lardum*, the fat of bacon); as the term ' waxy' is based upon its resemblance to wax. These names, thus understood, imply no theory as to the chemical composition of the morbid product, and are, therefore, not misleading, as the term amyloid unquestionably is.

Clinical History and Symptoms.—The clinical history of this disease has been carefully investigated ; and we are indebted to Dr. Grainger Stewart for insisting upon the fact that a copious flow of urine, of pale colour, and of low specific gravity, is one of the earliest and most constant symptoms of this form of degeneration. The urine at first may contain no albumen; but gradually it becomes albuminous, and often copiously albuminous. When a patient, whose strength has been reduced by a protracted and exhausting disease, begins to pass urine in large amount and of low density, so that his nights are disturbed by frequent calls to empty the bladder and to quench his thirst, we may anticipate that his kidneys are about to undergo the degenerative changes which we are now discussing. In the absence of sugar and albumen,

it may for a time be a question whether the disease is diabetes insipidus. The appearance of albumen points at once
to renal degeneration. The amount of urine secreted daily
usually ranges from 50 to 100 ounces or more. The colour
is pale, and the specific gravity varies from 1005 to 1015.
The urine may be clear, and deposit no sediment; so that
for days and even weeks together no tube-casts are visible;
and, in the earliest stages of the disease, tube-casts are
never numerous. In most cases, however, a light cloud
collects at the bottom of the conical glass; and in this
cloud we may find some small hyaline casts, some casts
finely granular, and occasionally some hyaline casts containing oil either in scattered globules or in cells. In
the advanced stage of the disease, when atrophy and
contraction of the kidney are in progress, the sediment in
the urine may be copious and dense; and it will be found
to contain numerous large-sized granular and hyaline casts,
exactly similar to those which I described as occurring when
the large white kidney has reached the third stage, and is
undergoing atrophy and contraction (see *ante*, Figs. 25,
26, and 27). As the disease advances, the patient's weakness increases; his breath is short on exertion; his coun-
tenance is pallid or sallow; the feet, ancles, and legs become
œdematous. In many cases, the liver and the spleen are
seen and felt to be more or less enlarged; and the abdomen is sometimes much distended by fluid. The disease
often has a very chronic course, extending over a period of
many months or even years. There may sometimes be a
temporary amendment; but in the majority of cases the
symptoms gradually become worse, until at length the
patient sinks, either from the direct effects of the renal
disease, or from one or other of the associated maladies.
The immediate cause of death may be dropsical effusion
within the chest, in the pleura or pericardium, or in both.
In other cases, the patient dies exhausted by diarrhœa, with
or without vomiting—the result, probably, of blood-deteri-

oration and the elimination of morbid materials through the mucous membrane of the alimentary canal. In some instances, the immediate cause of death is an attack of convulsions or coma. Cerebral symptoms of uræmic origin are, however, less frequent results of this than of other forms of chronic Bright's disease. The retina is rarely if ever affected. Inflammatory complications are of common occurrence. Of these, pneumonia is the most frequent ; next to this, inflammation of the serous membranes, especially of the pleura. The pericardium and the peritoneum are more rarely the seat of inflammation.

In many cases death results, not from the direct consequences of the renal degeneration, but from some associated constitutional disease or cachexia. Thus phthisis, or protracted suppuration, with or without disease of the bones or joints, or some form of constitutional syphilis, may be the immediate cause of death.

In the advanced stages of the disease there is extreme anæmia and pallor of the skin ; the blood contains much less than its due proportion of hæmoglobin and albumen, and in some cases an excess of urea.

Hæmorrhage from one or more mucous membranes, more especially from that of the nose, is an occasional occurrence in the advanced stages of the disease.

The Minute Anatomy and Pathology of the Lardaceous Kidney.—Most recent writers on renal pathology accept Virchow's theoretical interpretation of this disease, and assume that the first pathological change consists in thickening and degeneration of the walls of the minute arteries and Malpighian capillaries. In consequence of this degeneration, we are told, albuminous and fibrinous materials transude through the walls of the vessels, and infiltrate the tissues of the kidney ; and this is supposed to explain the structural changes in the gland. It has also been suggested that the copious secretion of urine in the early stage of the disease is a result and an indication of paralysis and

dilatation of the minute renal arteries, consequent on
degeneration of their walls. I find this theory inconsistent
with anatomical facts, and therefore I reject it entirely.
For a number of years I have most carefully studied the
condition of the renal blood-vessels in all forms and stages
of Bright's disease. My discovery of hypertrophy of the
muscular walls of the arteries was published in the year
1850 (*Med.-Chir. Trans.*, vol. xxxiii), some years before
the publication of Virchow's theory of amyloid degeneration. I soon learnt to distinguish muscular hypertrophy
from lardaceous and fatty degeneration of the arterial walls;
and I have carefully noted the microscopic appearances in
a large number of diseased kidneys. The result is that,
while I have not met with a single case in which thickening of the renal blood-vessels in any form was unassociated
with extensive changes in the secreting tissue of the kidney,
I have examined many kidneys in an advanced stage of
lardaceous disease with only incipient degeneration of the
blood-vessels. For example, I have before referred to one
case in which the two kidneys weighed twenty-eight ounces.
The patient at the time of his death was twenty-one years
of age. Since the age of three, he had suffered from
disease of the hip, with purulent discharge from several
openings about the joint. For ten or eleven years there
had been more or less dropsy, and for several months the
dropsy had been general. The clinical history, the character
of the urine, and the appearance of the kidneys were those
of a typical case of lardaceous degeneration of the kidney;
yet the Malpighian capillaries and the arteries in these
greatly enlarged, pale, and wax-like kidneys, after the
long duration of the symptoms, were only moderately
thickened. The gland had increased to nearly three times
its normal weight, while the vascular changes were in an
incipient stage. The increased size of the gland was not
such as could be explained by a mere infiltration of fibrinous material; but the enlargement was the result mainly

of an actual hypertrophy of the glandular tissue, precisely similar to that which I described in my last lecture as occurring in cases of the large white kidney. Most of the tubes were enlarged, and their epithelium was opaque from ' cloudy swelling.' In some, the cells were in a state of fatty degeneration ; and some tubes contained fibrinous coagula precisely similar to the large hyaline casts which had appeared in the urine during life. In some cases, more especially when the disease has gone on into the stage of atrophy and contraction, the gland cells have undergone more general and extensive degeneration. Numerous small yellow spots visible by the naked eye, when present, indicate the situation of tubes whose contents have undergone fatty degeneration (see *ante*, Fig. 18); while other tubes are filled with unorganised fibrine, which may sometimes be squeezed out of them in the form of large hyaline casts. The basement-membrane, both in the cortex and in the cones, sometimes appears thickened and hyaline ; and occasionally the tubular structure is rendered indistinct by an unorganised intertubular effusion. The Malpighian capillaries are thickened, opaque, glistening, and wax-like. (Fig. 30.) Some of the afferent arteries appear quite

Fig. 30.—Malpighian Capillaries, with Opaque Glistening Wax-like Walls. The Capsule somewhat thickened.—× 200.

normal, others are thickened by muscular hypertrophy ; but the greater number appear thick, more or less homogeneous, and wax-like ; their muscular structure being concealed apparently by an interstitial fibrinous infiltration. (See Fig. 31.) The straight arteries in the cones sometimes present the

same appearance of wax-like degeneration. If we now add
to the specimen a drop or two of diluted liquor potassæ
(one part of solution of potash in ten of water), the walls
of the waxy vessels are rendered transparent; so that the
red blood-corpuscles become visible through the thickened
Malpighian capillaries, and the muscular fibres of the
minute arteries are rendered quite distinct. The weak
alkali has a solvent action upon the infiltrated fibrinous
material, and thus to some extent brings into view the
normal structure of the arterial walls. The canals of the

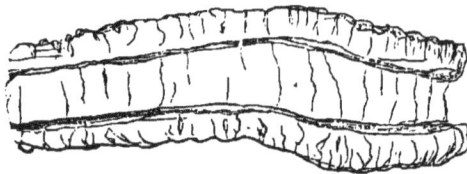

Fig. 31.—A Renal Artery with Lardaceous Infiltration of its Walls.—× 200.

afferent arteries may sometimes be seen irregularly dilated.
In some of the arterial canals, a collection of oil-globules
shows that the circulation has ceased sometime before death.
(Fig. 32.) Oil-globules may often be seen in the canals,

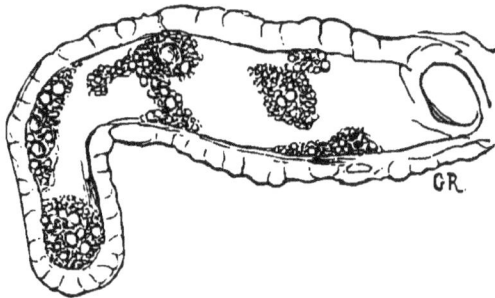

Fig. 32.—A Renal Artery, with Lardaceous Infiltration of its Walls and collections of Oil-
globules in the Canal, which is somewhat irregularly dilated.— × 200.

and in the walls of the Malpighian capillaries—less frequently within the intertubular capillaries. I have never

seen thickening of the walls of the intertubular capillaries. Thickening of the basement-membrane often gives an appearance of intertubular thickening. When fatty granulations are visible to the naked eye, their microscopic appearances are identical with those which I have before described in the granular fat kidney.

In my last lecture, I told you that the Malpighian capillaries and the walls of the minute arteries in some large white kidneys are thickened, homogeneous, and wax-like. The vessels in these cases have undergone precisely the same change as those in the lardaceous kidney. In fact, the large white kidney, which is found associated with chronic albuminuria following upon acute Bright's disease, and the lardaceous kidney, which occurs in connexion with chronic cachexia, have many points of contact and relationship ; and the two forms of disease merge into each other by imperceptible gradations. Chronic albuminuria is generally admitted to be one of the causes of lardaceous degeneration of the kidney.

All writers on the lardaceous kidney agree in stating that the primary cause of the renal degeneration is a morbid condition of the blood. In this opinion I concur; and I believe that some morbid material in the blood, acting upon the secreting tissues of the kidney, is the cause of the copious secretion of urine in the earlier stages of the disease, and of the glandular hypertrophy which gradually surpervenes during the progress of the malady. It is probable that the unknown morbid material in the blood has a diuretic influence upon the kidney, analogous to that which grape-sugar is known to have in cases of saccharine diabetes. We do not attempt to explain the diabetic urinary flux by the anatomical condition of the kidney ; neither can we thus explain the copious secretion of urine which precedes lardaceous degeneration. I say *precedes* the degeneration ; for my observations have convinced me that, while in the earlier stages of the disease the arterial walls are quite normal, at a later period they may become hypertrophied ; and later still, either with or without previous hypertrophy, they become infiltrated with fibrinous material, and assume the homogeneous waxy appearance. The Malpighian capillaries in every case of albuminuria have a more or less abnormal appearance. I told you in my second lecture that, after death from acute Bright's disease, the walls of the Malpighian capillaries are opaque and granular. In all cases of chronic albuminuria, the same capillaries are thickened ; and, when the transudation of albumen has been copious and long continued, the capillary walls always assume an opaque, glistening, wax-like appearance. The state of the Malpighian capillaries in the large white kidney is not distinguishable from that of the same vessels in the lardaceous kidney, either by microscopic examination or by any chemical test with which I am acquainted. The test which is usually applied is a weak solution of iodine. The thickened arteries and Malpighian capillaries take the reddish-brown stain much more deeply than the other tissues ; and the stained vessels consequently stand out in strong contrast with the pale, waxy, glandular structure. This iodine test, applied to the large white smooth kidney, often stains the vessels as deeply as when applied to the lardaceous kidney, and thus affords additional evidence of the close relationship between the two forms of disease. I cannot but think that too much importance has been attached to the iodine test, while too little attention has been given to the minute structural changes in the kidney, and their physiological interpretation.

Dr. Dickinson attributes so much influence to *suppuration* as causative of lardaceous degeneration, that he proposes to call the disease 'depurative,' using the word in a sense different from that in which it is commonly understood. Analyses of the morbid material, especially in the liver, have led him to the conclusion that it consists of partially dealkalised fibrine. He supposes that copious suppuration, by lessening the albumen and alkalies in the blood, causes a deposit of dealkalised fibrine in various tissues and organs. It seems not improbable that the diminished alkalescence of the waxy liver may be explained by the comparatively small proportion of alkaline blood which it contains. The liver-cells are infiltrated with, and in part replaced by, fibrinous material; while the vessels are compressed and anæmic. There is no apparent difference between the fibrinous exudation within the tubes of a lardaceous kidney and that which is found in the tubes of a large white kidney, the result of acute Bright's disease passing into a chronic form. Our knowledge of animal chemistry is as yet too elementary to enable us to give a chemical explanation of pathological changes so complex as those which we are now discussing; and, although the lardaceous disease is very frequently associated with suppuration, yet this is far from being constant. The renal degeneration occurs in only a small proportion of cases in which there has been profuse and protracted suppuration; and, on the other hand, the lardaceous form of renal degeneration not unfrequently occurs unassociated with a history of purulent discharges. Dr. Grainger Stewart states that in only six out of eighteen cases which he had himself carefully investigated, was there a history of suppuration. Obviously there is not that constant and close relationship between suppuration and lardaceous degeneration which justifies the application of the ambiguous term 'depurative' to this form of renal disease. It is probable that the deterioration of blood which results from long-continued profuse suppuration is due rather to the drain of albumen than to the loss of alkaline salts; and so it is intelligible that chronic suppuration and chronic albuminuria may bring about a similar condition of cachexia and malnutrition.

Hypertrophy of the Heart, unassociated with disease of

the valves or of the large arteries, rarely occurs, except in the cases which have passed on into the stage of atrophy. The explanation which I gave in my last lecture of the comparative infrequency of hypertrophy of the left ventricle in cases of large white kidney, is applicable here also. There is more of hydræmia than of uræmia associated with the lardaceous kidney; and the minute arteries offer little or no resistance to the passage of this watery blood. In addition, we have to take into consideration the fact that the walls of the minute arteries in various tissues and organs usually undergo degenerative changes, whereby their contractile power is impaired. The degeneration of the blood-vessels is of common occurrence in the mucous membrane of the intestines, and the change is rendered conspicuous by the iodine test. The absence of abnormal contraction in the terminal arteries explains the absence of the cardiac hypertrophy which ordinarily results from long-continued and excessive arterial resistance.

Diagnosis.—You will probably have inferred, from what I have said of the close relationship between the 'large white kidney' and the 'lardaceous kidney,' that it is often difficult to distinguish one from the other, and that the distinction has but little practical value. Even when you have the diseased organ before you, you may sometimes be in doubt whether to call it simply a 'large white kidney,' or to designate it 'lardaceous.' Obviously, then, it must sometimes be impossible to make the distinction during the lifetime of the patient. When there has been a copious secretion of urine, which for a time was free from albumen, but later has become copiously albuminous; when a copious secretion of pale albuminous urine, of low specific gravity, is associated with more or less general dropsy; when these symptoms have come on gradually and insidiously in a subject who has been suffering from an exhausting disease, such as phthisis, disease of the bones or joints, or cachexia resulting from cancer or constitutional syphilis, we may expect to find lardaceous degeneration. The probability of this will be much increased, if the liver or spleen, or both these organs, be found enlarged and indurated. A lardaceous kidney may sometimes attain a sufficient size to be palpable in the lumbar region. The tube-casts, when present, are essentially the same in the two forms of disease; and in particular the large hyaline and granular casts (Figs. 26 and 27), which appear in the advanced stages of both classes of cases, announce that atrophic changes are in progress; while the amount of sediment having these microscopic characters indicates the rate at which the destructive changes are proceeding.

Prognosis.—Although the history of this form of disease not unfrequently extends over a period of several years, as in the somewhat exceptional case to which I just now referred, yet the prognosis is, as a rule, very unfavourable —for the obvious reason, that not only is the renal disease often associated with serious structural change in other organs, but, resulting as it does from a grave constitutional cachexia, its causes are continually operating, and, as a rule, they are but little amenable to treatment. There may be occasional pauses in the progress of the disease, and even periods of temporary amendment; but the usual course of the malady is one of steady progress towards a fatal termination. The end is often hastened by an exhausting diarrhœa, by a copious dropsical or inflammatory effusion into the chest or abdomen, or by inflammation and sloughing of the dropsical legs.

Hæmaturia in Chronic Bright's Disease.—In conclusion, I wish to direct your attention to a possible source of fallacy resulting from the occasional appearance of blood-tinged urine in the advanced stage of this and of other forms of chronic Bright's disease. Dark-coloured, smoky, more or less blood-coloured urine, is of frequent occurrence in cases of acute Bright's disease; it is rare in the advanced stages of any of the three forms of chronic disease which

I have described, but it does occasionally happen ; and the appearance of hæmaturia might possibly mislead you in your estimate of the stage and gravity of the renal disease.

During the progress of the various forms of chronic Bright's disease, the walls of the Malpighian capillaries become thickened, and therefore probably less liable to be ruptured. In many cases, too, the muscular walls of the minute arteries are more or less hypertrophied; and the effect of this is to lessen the pressure upon the Malpighian capillaries and the risk of their rupture. This appears to be the explanation of the undoubted fact that the pale urine of low specific gravity which is secreted by kidneys in an advanced stage of degeneration, is rarely tinged with blood. This rule, however, is not without exceptions. In the dvanced stage of all forms of chronic Bright's disease, the blood becomes much deteriorated—partly, as we have seen, by the loss of its normal constituents; partly by the retention of urinary excreta. In some tissues, too, the minute arteries and capillaries may undergo degenerative changes which increase their liability to rupture. There is consequently a tendency to hæmorrhage from various mucous surfaces—from the nose, the lungs, the stomach and intestines, from the uterus, and sometimes from the mucous membrane of the bladder and the pelvis of the kidney. Hæmorrhage from the bladder, or from the pelvis of the kidney, may give the urine the dark colour and the blood-tinged appearance which it often has in cases of acute Bright's disease, when blood escapes from the substance of the kidney. You may come to a right judgment in these cases by a careful consideration of the past history, together with a close inspection of the urine. You will probably find that there are no blood-casts of the tubes, as there usually are when the substance of the kidney is the source of the bleeding. You may find some of those forms of tube-cast which point to the existence of chronic rather than recent acute disease : for instance, oily casts or large granular and large hyaline casts (Figs. 25, 26, and 27). You may also find that the urine, when, after standing for a time, it has deposited the blood, presents the pale colour which is indicative of chronic disease in an advanced stage. Many years ago, my friend Mr. James Salter sent me the notes of a case of *purpura* in which there had been profuse hæmaturia. The kidneys had been the seat of chronic Bright's disease; they were enlarged, anæmic, and had some cysts on their surface. The mucous membrane of the calyces, infundibula, and pelvis, was intensely congested and cho-colate-coloured with ecchymosis. There was a striking contrast and a sharp line of demarcation between the pale mamillæ and the dark ecchymosed calyces. It is probable that the absence of hæmorrhage into the substance of the kidney was due to the fact that the walls of the minute arteries and those of the Malpighian capillaries had become thickened during the progress of the chronic degeneration of the kidney which had long preceded the appearance of the purpura. This case affords a good illustration of the fact that hæmaturia, the result of blood-deterioration, may have its source in the pelvis, and not in the substance of the kidney. A microscopic examination of this patient's urine had discovered no tube-casts. None of the blood had been moulded within the uriniferous tubes, because none had escaped from the Malpighian capillaries.

LECTURE VI.

Albuminuria not associated with what is commonly understood as Bright's Disease—1. Passive Congestion of the Kidney with Albuminuria, the Result of Impeded Venous Circulation—2. Albuminuria and Hæmaturia from Embolism in the Minute Blood-Vessels of the Kidney—3. Puerperal' Albuminuria—Four Classes of Cases—4. Atrophy and Suppurative Inflammation of the Kidney from Retention of Urine—Symptoms and Diagnosis—5. Acute Cystitis resembling Acute Bright's Disease—Symptoms, Diagnosis, and Treatment.

BEFORE I go on to discuss the treatment of Bright's disease, which I shall do in my next lecture, I wish to direct your attention to certain cases of albuminuria resulting from various causes, but not associated with what is commonly understood as Bright's disease. My object in referring to these cases now, is to give you some hints which may assist you to distinguish them from each other and from cases of actual Bright's disease. My remarks upon these cases of albuminuria not associated with Bright's disease will have reference mainly to pathology and diagnosis, with only an occasional suggestion on the important subject of treatment.

1. *Passive Congestion of the Kidney with Albuminuria, the Result of Impeded Venous Circulation.*—There is a class of cases in which albuminuria results from passive congestion of the kidney consequent on some impediment to the return of blood through the systemic veins. The causes of this impediment are diverse. Valvular disease of the heart is one of the most frequent of them. Degeneration and consequent weakness of the muscular walls of the heart is a not uncommon cause. The impediment may originate in the lungs, as a result of emphysema with bronchitis, of extensive pneumonic consolidation, or of compres-

sion of one or both lungs by a copious liquid effusion into the pleura. Again, a dropsical effusion in the cavity of the peritoneum, the result of cirrhosis or other obstructive disease of the liver, may so compres̓s the vena cava and impede the return of venous blood, as to cause passive renal congestion and albuminuria. In the advanced stages of pregnancy, the pressure of the uterus on the veins may cause passive renal congestion and albuminuria; but I shall presently refer more particularly to albuminuria in connexion with pregnancy. In general, the diagnosis of each of these causes of impeded circulation is not difficult. Then, as to the effect upon the urine and the kidney: the urine becomes scanty in proportion to the degree of venous congestion and the consequent tardiness of the blood-stream through the kidney. The secretion is usually high-coloured, of high specific gravity, often turbid with urates, and more or less impregnated with albumen.

The mechanism of albuminuria consequent on the passive engorgement of the kidney which results from an impeded return of blood through the veins, may be illustrated by reference to Fig. 1, Lecture 1. When, in consequence of an obstruction at the heart, the systemic veins become overfull, the distension of the renal vein, acting backwards through the intertubular capillaries, causes engorgement of the Malpighian capillaries, and a consequent transudation of serum through their walls. This serous transudation, mingling with the urine, renders it albuminous. Small hyaline and granular casts may be seen when the turbid urine has been cleared by warmth or by dilution with water. The Malpighian capillaries are sometimes ruptured by over-distension ; the urine is then blood-tinged, and blood-casts are visible. The secondary character of the renal complication is usually apparent from the history of these cases ; and the diagnosis may sometimes be confirmed by the fact that, when the circulation has been relieved by rest in bed, by hydragogues, by puncturing the legs, or by tapping the abdomen, the albumen disappears from the urine; to return, perhaps, when the circulation again becomes more embarrassed within the chest or by the reaccumulation of liquid in the peritoneum. I have seen this happen again and again during the progress of the same case.

The first effect of passive congestion upon the kidney is to cause more or less enlargement with some induration of the gland. The ultimate result of long-continued congestion is atrophy and contraction, the surface of the kidney becoming uneven and finely granular as the wasting process goes on. The explanation of the phenomena is not difficult. For the due performance of its secretory function, and for the maintenance of its nutrition, it is essential that the blood move freely through the gland. An impeded return of blood through the veins involves as a necessary consequence a partial blood-stasis, and, as a result of this, a scanty secretion of urine, with impaired nutrition and atrophy of the gland. On microscopic examination of the kidneys, some of the tubes may be seen to be opaque with disintegrated epithelium and fibrine, some denuded, and in various stages of atrophy and contraction. When atrophy of the kidney has been a result of passive congestion consequent on a mechanical hindrance to the circulation, I have never found the muscular walls of the minute renal arteries hypertrophied.

II. *Albuminuria and Hæmaturia from Embolism in the Minute Blood-Vessels of the Kidney.*—There is yet another mode in which valvular disease of the heart may for a time render the urine albuminous and even bloody. You are aware that, when one of the valves of the heart has its surface roughened by inflammation or by senile degenerative changes, a very common result is a deposit of fibrine upon the roughened surface; and further, that these fibrinous deposits, having no organic union with the valve beneath, are very liable to become detached by the current of blood, and then to obstruct the vessels in any organ to which they may chance to be conveyed. One result of this mechanical plugging of blood-vessels is the formation of so-called 'fibrinous deposits' in the kidney. The portion of kidney which is the seat of recent obstruction is raised above the level of the surrounding renal tissue; it is firm, anæmic, and of a yellowish white colour, with an intensely red injected margin. The older deposits are softer than the surrounding tissue, appear shrunk and depressed, and have not the red margin. In a still more advanced stage, the appearance of a deposit entirely passes away, and a depressed cicatrix is left.

On a microscopical examination of a recent fibrinous patch, the tubes in the seat of the deposit appear opaque from containing fibrinous coagula. Some tubes contain oil. Many of the intertubular capillaries contain fibrinous coagula; while others contain oil-globules, which are clustered in the form of rings surrounding the tubes. Granular coagula and oil-globules may also be seen in some of the Malpighian capillaries and the afferent arteries. The coagula in the vessels are more clearly seen after the tissues have been rendered transparent by dilute acetic acid. In the red vascular zone which surrounds the recent deposit, the Malpighian and the intertubular capillaries are seen to be injected and gorged with blood. The probable explanation of the phenomena is, that a soft fibrinous mass from a cardiac valve is arrested in a small branch of a renal artery; and there it becomes disintegrated, and the fragments are carried on into the intertubular capillaries. The circulation through these vessels is arrested; there is, consequently, a backward engorgement of the Malpighian capillaries, and an albuminous and fibrinous effusion into the uriniferous tubes. The capillaries at the margin of the obstructed patch are greatly distended by the diverted current of blood, and in consequence the urine may become albuminous and even blood-tinged; hyaline and blood-casts being visible under the microscope. Subsequently, the exuda-

tion into the tubes, the epithelium of the tubes, and the fibrinous coagula within the obstructed blood-vessels, undergo a
fatty transformation, and all trace of the normal structure
disappears. The fatty matter at length becomes absorbed,
and a depressed cicatrix remains on the surface of the
kidney. Two or more pale fibrinous patches of different
dates may sometimes be found in the same kidney. In
some instances, the fatty matter which results from the
transformation of the tissues and the fibrinous coagula does
not become absorbed, but remains encysted. This is the
explanation of the cysts which are sometimes found filled
with a thick dark liquid, composed of oil, free and in cells,
with which often plates of cholesterine are mingled.

The diagnosis of embolism in the renal vessels is usually
more or less uncertain. We may suspect the occurrence
when, with the physical signs of aortic or mitral disease,
without great impediment of the general circulation, the
urine suddenly becomes albuminous or bloody. In some
cases, extensive embolism in one or both kidneys has been
attended with severe lumbar pains, a scanty secretion of
urine, and vomiting; but, when the obstructed portions of
kidney are small, there may be no symptoms to indicate the
occurrence of renal embolism.

III. *Puerperal Albuminuria.*—Since the time when Dr.
Lever, in the 'Guy's Hospital Reports' (1843), published the
fact that puerperal convulsions are in a large proportion of
cases associated with albuminuria, the subject of albuminuria
in connexion with pregnancy has excited much interest.
Later observations have established the essential accuracy of
Dr. Lever's observations; but they have also shown that the
connexion between puerperal convulsions and albuminuria is
not constant. Convulsions may occur without albuminuria,
and, on the other hand, albuminuria in pregnant women
may be unassociated with convulsions. The few remarks
which I propose to address to you on this subject will have
reference mainly to the subject of puerperal albuminuria,

and only incidentally to the association of puerperal convulsions with albuminuria. With reference to the subject
of puerperal convulsions, I advise you to read carefully Dr.
Barnes's very able and interesting Lumleian Lectures 'On
the Convulsive Diseases of Women.' (Published in the
'British Medical Journal,' 1873.) I have seen more or less
of a considerable number of cases of albuminuria associated
with pregnancy; and, looking over my notes of these cases,
I find that they arrange themselves in four classes, each
having in some respects a different history and pathology.

1. Women known to be suffering from chronic Bright's
disease may become pregnant, pass through all the stages
of pregnancy and parturition, and even suckle their infants,
without accident or complication. A lady whom I saw
some years since with my friend Dr. S. H. Steel of Aber-
gavenny, while suffering from chronic Bright's disease
supervening upon an acute attack which resulted from
exposure to cold, twice became pregnant, each time had an
uncomplicated labour, and gave birth to two healthy children, each of which she suckled for about a year, not
only without detriment, but apparently with benefit to her
health. In 1849 I first saw with the late Dr. Tanner a woman
who during her tenth pregnancy had general dropsy with
albuminuria. The labour was natural, the child was healthy,
and nursed for a time. After that, while the urine continued
to be albuminous, she twice became pregnant, and had uncomplicated labours. She ultimately died with contracted
granular fat kidneys. It is probable that the existence of
albuminuria from any cause increases the risk of puerperal
convulsions at the time when to the exalted reflex excita-
bility of the nervous system in the parturient woman there
is superadded the disturbing element of violent and general
muscular contraction; but cases like those which I have
cited show that, when there is a free secretion of urine, with
absence of uræmic symptoms, labour may be unattended
with any serious complication.

2. In a second class of cases, during the latter months. of pregnancy, there is more or less general œdema, with headache and other nervous symptoms, not unfrequently culminating in convulsions, which may recur again and again. The urine is scanty, high-coloured, often turbid with urates, of high specific gravity, and contains a large amount of albumen. On microscopic examination it is found to contain small hyaline casts, with a few granular casts, but few or no epithelial casts. After delivery the urine quickly becomes copious, of pale colour, of low specific gravity, and within forty-eight hours the albumen may have entirely disappeared. Such a case I saw in the year 1857 with Dr. Greenhalgh and Mr. Peter Marshall. The most probable explanation of this class of cases is, that the pressure of the gravid uterus on the vena cava causes gradually increasing passive engorgement of the kidney, albuminuria, a scanty secretion of urine, dropsy, and at length uræmic convulsions. The rapid disappearance of the albuminuria and the other symptoms after the emptying of the uterus is explicable on no other theory than that of passive renal congestion consequent on mechanical pressure. Cases of this class are more common in primiparæ, and probably for the reason that in first pregnancies the abdominal walls are less yielding; there is, therefore, greater tension and greater pressure on the large venous trunks than during subsequent pregnancies, when the abdominal walls are more flaccid.

3. There is a third class of cases, in which the theory of mechanical pressure is not admissible. I allude to those cases in which albuminuria comes on at an early period of pregnancy, before the uterus has attained sufficient size and weight to interfere mechanically with the circulation through the kidney. In these cases there is sometimes evidence of acute desquamative nephritis. The urine is not only scanty and highly albuminous, but often blood-tinged, and contains epithelial and blood-casts. Such a case I have in the

hospital at the present time (March ·1873). When albuminuria sets in during the progress of pregnancy, it is very
apt to lead on to convulsions, to retinal hæmorrhage and
albuminuric retinitis, with serious defect of vision. The
renal symptoms may gradually pass away after delivery;
if so, they may or may not return with the next pregnancy.
In other cases the albuminuria is persistent; the urine is
of pale colour, of low specific gravity, and deposits small
hyaline and granular casts. Ultimately, uræmic symptoms occur; and, after death, the kidneys are found either
contracted and granular or large and pale. A painful case
of this kind I saw not long since with Mr. Cadge of Nor-
wich. The most probable explanation of this class of cases
is that which refers the renal disease to some previous
blood-change. Obviously, pregnant women are exposed to
the ordinary exciting causes of renal disease; and acute
Bright's disease originating during pregnancy may result
from exposure to cold and wet, from excessive eating and
drinking, or from some zymotic blood-poison. But, in
addition to these more common causes of albuminuria, it
is probable that, connected with the evolution of the
uterus and the development and growth of the fœtus, there
may sometimes be associated abnormal blood-changes, resulting in renal disease with albuminuria. In addition to
other indications of the occasional occurrence of morbid
states of blood in pregnant women, I may refer to those
cases in which puerperal chorea is associated with acute
endocarditis and fibrinous deposits in the mitral or aortic
valves; the chorea being, in all probability, a result of
capillary embolism in the region of one or both corpora
striata.

4. There is yet a fourth class of cases, of which I have
seen and noted several examples. I refer now to cases in
which albuminuria and other symptoms of renal disease
appear for the first time soon after delivery. Within a day
or two after delivery, or after an interval of several days,

sometimes after imprudent exposure to cold, a rigor occurs, and is followed by febrile symptoms. The urine is soon found to be scanty, with all the characters indicative of acute desquamative nephritis. There may be general dropsy, with or without uræmic nervous symptoms, such as headache and convulsions. The renal symptoms, after a period varying from a few weeks to several months, may gradually and entirely pass away; or the disease may become chronic, and result in a large white kidney. In such cases as this, the renal symptoms may with confidence be referred to the blood-contamination consequent on the absorption of morbid materials from the interior of the uterus after parturition. These cases are pathologically allied to, and sometimes associated with a form of septicæ-mic puerperal fever. An interesting case of this kind, about which I was consulted, has been published by Messrs. Mel-land and Windsor, of Manchester, in the ' British Medical Journal,' September 12th, 1847. Here, too, I would suggest the probability that, when the fœtus dies and is retained *in utero* until decomposition is commenced, there may some-times be an absorption of foul gases and liquids, which in one case may give rise to the phenomena of *ante partum* phlegmasia dolens, as in a case which we have recently had in the hospital; and, in other instances, acute desquamative nephritis may result from this source of blood-infection.

It will be obvious that the distinction between the four classes of cases of albuminuria in connexion with pregnancy, which I have here briefly indicated, is of considerable prac-tical importance, inasmuch as upon an exact diagnosis depends, not only the prognosis but the treatment of each case of puerperal albuminuria.

One hint I may give you with reference to the expediency of allowing a woman suffering from albuminuria to nurse her infant. The case of Dr. Steel's patient, to which I just now referred, shows that, when there are no symptoms of blood-poisoning, the mere fact of albuminuria does not prevent a woman from being a good nurse; but, on the other hand, when albuminuria is a result of a recent blood-infection, the mother's milk may become contaminated, and act as a poison to her infant. In one case about which I was consulted, a lady had acute renal disease, resulting probably from *post partum* absorption from the interior of the uterus. She recovered after a long illness, complicated with pelvic cellulitis and abscess; but her infant, after taking the breast for five weeks, became feverish, and died with symptoms of pyæmia at the age of six weeks. It seemed probable that the infant's illness was a result of infection through the milk; the infection being not uræmic, but septicæmic—a consequence of the absorption of noxious uterine discharges. In cases similar to this, the mother should not be permitted to nurse her infant.

IV. *Atrophy and Suppurative Inflammation of the Kidney from Retention of Urine.*—The effect upon the kidney of retention of urine varies according to the nature and seat of the impediment. It differs, too, according as the obstruction occurs gradually or suddenly. One of the most frequent causes of renal disease consequent upon retention of urine is stricture of the urethra. The urinary organs behind the stricture undergo changes of structure in proportion to the degree and the duration of the obstruction. The canal of the urethra on the vesical side of the stricture becomes dilated; its mucous membrane is frequently inflamed, and secretes pus. The muscular coats of the bladder become thickened by hypertrophy, and its mucous membrane often inflamed and sacculated. The obstruction then affects the ureters, one or both of which may have their canals dilated and their walls thickened; and at length the natural cavities of the kidney—the pelvis, infundibula, and calyces—undergo the same process of dilatation. The medullary cones become flattened out by the pressure of the retained urine. The cortical substance of the gland is expanded, and presents lobed bulgings on its surface, which correspond with the original lobes of the embryo kidney. The glandular tissue is squeezed between the distended interior cavity and the fibrous investing capsule; and the intertubular capillaries are compressed by dilated tubes. Thus the circulation is impeded, and the result is atrophy of the gland, which may by degrees be converted into a membranous cyst, all traces of glandular structure being lost.

It is but seldom that the kidney undergoes much dilatation without the occurrence of other structural changes. The mucous membrane of the dilated pelvis often presents irregular inflamed patches, and secretes a purulent liquid; and the apices of the medullary cones are frequently ulcerated. Then, as the mischief extends, inflammatory deposits occur in the substance of the kidney, and numerous small abscesses are scattered through the cortex. One or more of the abscesses on the surface may burst through the capsule, and then the kidney may be found imbedded in pus.

When the retention of urine is the result of stricture, or enlarged prostate, or calculus with thickening of the walls of the bladder, or of atony of the muscular coats, both kidneys are usually affected simultaneously, but in different degrees; but when one ureter is obstructed by a calculus, or by a cancerous growth, the structural changes are limited to the corresponding kidney.

The explanation of these changes is not difficult. The secreted urine is unable to escape in consequence of the obstruction in front; there is, therefore, an accumulation, first in the ureter and pelvis of the kidney, and later within the uriniferous tubes. The tubes become distended by the retention of their own secretion, just as some tubes in the small red granular kidney, having lost their lining of gland-cells, but continuing to secrete a serous liquid, become distended and dilated into cysts. The epithelial lining of the straight tubes is disintegrated and destroyed by the pressure of the retained urine; and at length some of the

tubes in the cortex, whose basement membrane is more
delicate than that of the cones, give way, and allow their
contents to become infiltrated amongst the intertubular
capillaries. The infiltration of acid urine may cause the
immediate formation of coagula within the capillaries, and,
as a consequence of localised capillary and venous obstruction, irregular atrophic puckerings of the gland occur,
somewhat similar to those which result from capillary embolism, to which I have before referred. But the escape
of urine through the broken walls of the tubes may excite
suppurative inflammation and abscess. A rapid cell-forma-
tion takes place between the tubes ; and soon the glandular
structure is disintegrated, and replaced by inflammatory
products. The changes within and between the uriniferous
tubes are a miniature representation of what happens on a
larger scale when a distended urethra gives way behind a
stricture, and a perineal abscess results from the infiltration
of urine into the submucous tissues. Bear in mind that the
changes within the substance of the kidney are due to the
retention and accumulation of the newly secreted acid urine
within the tubes, and not to the regurgitation of fœtid
ammoniacal urine, as has recently been suggested. Without doubt, the urine in the bladder in cases of old stricture,
vesical calculus, and cystitis, is often fœtid from decomposition ; but the regurgitation of such urine into the uriniferous tubes is a physical impossibility. If you have ever
attempted to inject the tubes from the pelvis of the kidney,
you will have found the task a very difficult one, in consequence of the resistance offered by the liquid and solid contents of the closed tubes. Obviously, then, during life, while
streams of secreted urine are perpetually flowing through
the tubes, it is impossible that urine from without can re-
gurgitate into them. The tubes are dilated, and some of
them ultimately ruptured, by the retention and accumulation
of their contents, and not by the regurgitation of urine from
the pelvis of the kidney. Then intertubular coagula and

suppuration result from the infiltration of urine amongst the intertubular capillaries and veins.

The *Symptoms* of renal disease consequent upon an impeded escape of urine are usually more or less masked by the diseased condition of other parts of the urinary organs. The mucous membrane of the bladder in cases of stricture, vesical calculus, or enlarged prostate, usually secretes pus; and there are no means by which this can be distinguished from matter derived from a suppurating kidney. For the suppurative process in the kidney rapidly destroys the tubular structure of the organ; the pus, therefore, is not moulded within the tubes, and there is no microscopic evidence of the renal origin of the pus. Chemistry, again, affords no more assistance than the microscope. The urine, which contains pus, is always albuminous. The coagulability of the urine by heat and acid is, therefore, no indication that the kidneys are implicated, except when the degree of coagulability is out of proportion to the amount of pus, and it may be blood, mingled with the urine. Chemical analysis affords little practical aid in estimating the efficiency of the kidney. The urine is usually foetid and alkaline, and much of the urea is decomposed into carbonate of ammonia while the urine is still in the bladder. A low specific gravity of the urine with a scanty secretion would be a suspicious condition, and especially so when associated with indications of uræmia, such as drowsiness, headache, vomiting, and a brown dry tongue, with an excess of urea in the blood. Pain and tenderness in the region of one or both kidneys may be severe when, with sudden retention of urine, there is great distension of the cavity of the kidney; but, in cases of long continued and slowly increasing obstruction, these symptoms bear no proportion to the amount of structural change in the kidneys; and it sometimes happens that the first indication of serious renal disease is afforded by the occurrence of alarming symptoms of uræmic poisoning, quickly passing on to fatal typhoid collapse and coma, with a low temperature. The cases are few in which the kidneys are sufficiently enlarged by distension and expansion to form a palpable tumour in the lumbar region.

v. *Acute Cystitis simulating Acute Bright's Disease.*— It has happened to me to meet with a considerable number of cases of acute inflammation of the mucous membrane of the bladder, unconnected with stone, stricture, or gonorrhœa, which, in consequence of the urine being blood-tinged and albuminous, have been mistaken for cases of acute Bright's disease. Therefore, before I proceed to describe the treatment of Bright's disease, I think it well to point out to you the distinctive features of acute cystitis. Remember that I exclude from our present consideration such obvious and common cases as cystitis from stone, stricture, retention of urine, and gonorrhœa; and I refer to cystitis not excited by any obvious mechanical cause. In a large proportion of cases—in six out of twelve of which I have notes—the disease directly followed, and was probably caused by, a chill. In four cases, dyspepsia with rheumatic or gouty symptoms had preceded the cystitis. In one, the disease came on after feasting with excess of wine. In one case—that of a physician—the symptoms commenced within a few hours after he had been impressed by a peculiar odour from the throat of a boy whom he was attending with a low form of scarlet fever. It is probable that in all these patients the immediate cause of the cystitis was some irritating material in the urine. In the cases last mentioned, some poisonous product may have entered the circulation, and passed out through the kidneys without exciting disease in them, but setting up inflammation in the bladder. It is difficult to explain or to understand how it happens that exposure to cold should in one person excite acute cystitis, and in another acute desquamative nephritis. Out of the twelve cases of acute cystitis which I have noted and tabulated, eight were males, and four females. The ages ranged from seventeen in a female to sixty-nine in a male. In some cases there has been more or less of vesical irritation for a few days before the acute attack, but in most instances the onset has been sudden and severe. The chief symptoms are frequent micturition, with more or less of uneasiness or pain in the region of the bladder. The calls to pass urine may occur every half-hour, or even oftener ; and micturition is usually attended with an increase of pain and a sense of scalding in the neck of the bladder. The vesical irritation is increased by exercise, by exposure to cold, and by alcoholic liquors. The urine quickly becomes turbid with puriform mucus, and often blood-tinged. In one case there was a puriform discharge from the urethra as well as from the bladder, though the disease was certainly not the result of gonorrhœa. Usually the urine has an acid reaction ; but, if there be much admixture of blood, the acidity is lessened by the alkali of the blood. It contains an abundance of albumen, partly derived from the puriform secretion, partly from the blood. On a microscopic examination, pus-cells and blood-corpuscles are seen in abundance, but no tube-casts. Although the local symptoms are distressing, there is little or no constitutional disturbance. The nights are disturbed by frequent calls to micturate, and the broken rest is attended with a sense of fatigue and nervous exhaustion ; but there is little fever, and no vomiting. If, within a few days from the onset of the symptoms, the patient be subjected to appropriate treatment, the disease usually subsides as rapidly as it came on. If, on the other hand, the symptoms be negligently or erroneously treated, the disease may become chronic, and cause prolonged and severe suffering. The urine becomes alkaline, ammoniacal, and fœtid ; there are perpetual pain and annoyance ; and ultimately the disease may extend backwards through the ureters to the kidneys, and so set up a fatal pyelo-nephritis.

Diagnosis.—The distinction between acute cystitis and acute Bright's disease is sufficiently obvious if you bear in mind that the local symptoms are all referable to the bladder, while dropsy, vomiting, and other renal symptoms are absent. The urine is usually secreted in normal quantity, and of normal specific gravity. It is albuminous only in direct proportion to the amount of blood and pus which it contains, and the most careful microscopic examination discovers no tube-casts.

Treatment of Acute Cystitis.—In the treatment of this form of cystitis, you have to bear in mind that the inflammation of the lining membrane of the bladder is kept up and increased by contact of the acid and irritating urine. The main object of treatment is to diminish as much as possible the irritating qualities of the urine; with this object in view, the patient must be confined to bed or to a sofa in a warm room, and be kept on liquid food without stimulants. If there be no special reason in the peculiarity of the patient's stomach to forbid it, milk, cold or tepid, may serve as meat and drink—as a rule, milk is easily digested, and, at the same time, it acts as a diluent. If milk disagree, soup, beef-tea, mutton- or chicken-broth may be given, with the addition of some farinaceous pudding, and any simple diluent drink. Pure cold water is as efficacious, though not so pleasant, as aërated or Seltzer water. In addition, the urine is to be kept neutralised by citrate of potash, given every six hours. A warm hip-bath should be given night and morning; a dose of morphia at bedtime, to allay irritation and procure sleep; and an occasional seidlitz powder as a laxative if necessary. Under this plan of treatment, the acute symptoms usually subside with great rapidity; the pain and irritation pass away, and the urine regains its normal characters. After a few days, the citrate of potash may be discontinued, and the tincture of perchloride of iron given in twenty minim doses two or three times a day after food. So long as any mucus appears in the urine, even though the local uneasiness may have passed away, the patient should be kept under observation

K

and treatment. If a chronic catarrhal condition of bladder remain after the acute symptoms have subsided, copaiba balsam often effects a rapid and complete cure. One capsule may be given an hour or two after food three times a day; and, if the stomach will bear it, the dose may be increased until six, and even nine capsules are taken in three doses in the twenty-four hours. In one case which had been of a year's duration, the urine being turbid with blood and pus, and smaller doses of copaiba having failed to cure, at the suggestion of Sir William Fergusson, who saw the patient with me, the dose was increased to three capsules three times a day, and the result was a complete and permanent cure within six weeks; the urine at the end of the treatment being of a natural sherry colour, transparent, and without a trace of mucus. Quite recently I have seen a young lady in whom acute cystitis from cold had left a vesical catarrh after a period of six weeks. I prescribed one capsule three times a day, and, in less than a week, the urine was entirely free from mucus, and the cure was complete. I have seen equally good results in other cases. It is probable that the remedy, being excreted by the kidneys, has a local curative action on the mucous membrane of the bladder. The copaiba sometimes brings out a transient erythematous rash on the skin. When this occurs, the dose of the copaiba must be lessened, or it may be necessary to discontinue the medicine for a time, and then perhaps to resume it in smaller doses.

LECTURE VII.

THE TREATMENT OF ACUTE AND CHRONIC BRIGHT'S DISEASE.

I HAVE endeavoured to prove to you that the various forms of Bright's disease are physiological results of the excretory function of the kidney. In accordance with the now fashionable language of modern biologists, we may say that the different forms of Bright's disease are results of a physiological process of evolution, and not of new pathological creations. The kidney is one of the main channels by which effete and noxious materials are cast out of the circulation. During the process of excreting abnormal products, the tissues of the kidney—primarily the gland-cells, secondarily the blood-vessels and the connective tissue—undergo structural changes. It follows from this interpretation of the pathological changes in the kidney, that a leading principle of treatment is to lessen as much as possible the excretory work of the kidney by instructing the patient to avoid the exciting causes of his malady, by a carefully regulated diet, and by such remedial agencies as experience has proved to be beneficial.

In all cases of *acute Bright's disease*, whatever may have been the exciting cause, rest in bed and in a room of moderate uniform temperature, well ventilated, but without chilling drafts of cold air, is an essential part of the treatment. In a large proportion of cases, rest in bed, with a scanty diet and a liberal use of diluent drinks, will suffice for the cure. A convincing proof and illustration of the effect of exercise, food, and cold, upon the amount of albumen in the

urine, is afforded by the fact that, in most cases of albumi-
nurie, the urine passed after rest in bed and before breakfast contains much less albumen than that which is secreted
after exercise in the open air and after an ordinary meal.

The diet may consist of milk alone, if milk do not disagree, as it does with some patients. Milk is especially
suitable for children; and it serves both for meat and
drink, so that no other food or liquid need be taken. It
may be taken cold or tepid, from half a pint to a pint at a
time. An adult will take sometimes as much as a gallon
in the twenty-four hours. Children will take less, in proportion to their ages. If the cream disagree, causing
heartburn, diarrhœa, headache, or other symptoms of dyspepsia, the milk may be given skimmed. One reason,
amongst others, for giving the milk as a rule unskimmed
—that is, with the cream—is, that constipation, which is
one of the most troublesome results of an exclusively milk
diet, is to some extent obviated by the cream in the unskimmed milk. The advantage of milk as a main article
of diet is that, as a rule, it is easy of digestion; and that,
taken freely, it supplies an abundance of liquid, which, by
its diluent action, has a diuretic influence, and so favours
the removal of the dropsy. There are some patients with
whom, unfortunately, milk in any form, even in small
quantities, so decidedly disagrees, that we have to find a
substitute in beef-tea, chicken- veal- or mutton-broth, with
an egg or two, and some farinaceous addition, such as
barley-water, arrowroot, rice, or sago, or a small quantity
of bread. Under this regimen, adopted and rigidly carried
out at the very commencement of acute Bright's disease,
the urine soon becomes copious, while the albumen diminishes and gradually disappears, and the dropsy quickly
passes away.

In my second lecture, I gave you the physiological explanation of the copious flow of urine which usually occurs
during convalescence from acute Bright's disease, and espe-

cially when there has been a copious dropsical effusion. This abundant flow of urine usually occurs without aid from diuretic drugs, or, indeed, from drugs of any kind. Stimulating diuretics, such as squills, cantharides, or turpentine, would be injurious by increasing congestion of the kidney. The best diuretics in cases of acute Bright's disease are those means which tend to lessen congestion of the kidney; such as dry cupping or hot poultices or fomentations over the loins, warm baths, and a free use of diluent drinks, one of the pleasantest and most efficacious being the ' imperial drink,' made with cream of tartar and lemon.

In the later stages of an acute attack, when the dropsy has disappeared, the urine being normal in colour, quantity, and specific gravity, but still more or less albuminous, a too free use of diluent drinks may be injurious by diluting the gastric secretions and so impeding digestion ; and by exciting an increased secretion of urine, which carries with it an exhausting discharge of albumen.

Warm baths are particularly useful in the early stage of an acute attack, and more especially when exposure to cold has been the exciting cause of the renal disease. A warm water bath, at a temperature of 98 or 100 deg., may be given every night during the first few days of an acute attack ; or a hot-air lamp-bath ; or, what I believe in most cases to be still more efficacious, a wet sheet and blanket bath. A sheet is wrung out of warm water ; and the patient, either naked or covered only by his shirt, is enveloped in the wet sheet up to the neck. Then three or four dry blankets are closely folded over the wet sheet. He may remain thus packed from two to four or six hours, or even longer. Recently, a boy in the hospital with acute renal disease and almost complete suppression of urine, consequent on scarlet fever, was kept packed incessantly for four days without serious discomfort, and with great relief from very distressing and alarming symptoms. When he left the hospital, all traces of his malady had disappeared. If the packing be long continued, the sheet has to be rewetted as soon as it becomes dry. The evaporation and consequent drying of the sheet will be slow in proportion to the closeness of the blanket packing. If the outer blanket be covered by a mackintosh cloth, the sheet remains wet for a much longer time than when no waterproof covering is used; but patients often complain of a feeling of oppression when surrounded by the impervious mackintosh. The advantage of the blanket-bath over a warm water or hot air bath, is, that it requires no special apparatus, that the diaphoretic action may be more prolonged, and that in most cases it is more agreeable to the patient. The hot air bath not unfrequently causes an unpleasant throbbing in the head, or a feeling of exhaustion and even faintness. When the wet pack is removed, the patient should be quickly rubbed dry, and enveloped in dry blankets.

The diaphoretic action of any form of warm bath is assisted by copious libations of simple diluent drinks, and it may also be aided by the internal administration of the solution of acetate of ammonia. It has been objected on theoretical grounds, that to promote perspiration in these cases is injurious by diverting to the skin the water which is required to wash out the uriniferous tubes. To meet this objection, you have only to bear in mind that the dropsical patient is oppressed by an excess of water, which has been poured into the areolar tissue and the serous cavities, in consequence of defective urinary secretion resulting from the inflammatory engorgement and obstruction of the kidneys. If, therefore, by the relaxing effect of external warmth, you divert a large amount of blood to the surface, you thereby lessen the congestion of the kidney, increase the freedom of the renal circulation, and so favour the occurrence of that copious secretion of urine which is one of the surest signs of satisfactory progress, and by which the uriniferous tubes may be effectually flushed and cleansed. Whatever fluid is lost by perspiration, may be quickly restored by the liberal use of diluent drinks, which again assist the secretory activity of both the skin and the kidneys. In fact, one of the main objects of treatment is to increase the freedom of the circulation, more especially through the kidneys; and thus to get rid of the excess of stagnant water which has accumulated in consequence of defective action of the skin and kidneys.

Purgatives may be usefully combined with other means for lessening dropsical effusion. In ordinary cases of acute Bright's disease, I do not, as a rule, advise the frequent employment of drastic purgatives. I reserve this method of treatment for cases in which there is an excessive and increasing dropsical effusion which does not yield to other means of cure, but more especially for cases in which cerebral symptoms, the result of uræmia, are either present or apparently impending. Dr. Abercrombie and others, who wrote on brain-disease before Dr. Bright's discovery had led on to our present knowledge of uræmic nervous symptoms, published cases of cerebral disease which they took to be of an inflammatory or an apoplectic character, in which the most strikingly beneficial results were obtained by free purging. Dr. Abercrombie, in discussing the treatment of inflammatory affections of the brain, states ('Pathological and Practical Researches on Diseases of the Brain,' 3rd edition, p. 153) that, according to his own experience, 'more recoveries from head-affections of the most alarming aspect take place under the use of very strong purgatives than under any other mode of treatment.' Our more recent experience is entirely confirmatory of Dr. Abercrombie's statement; but our improved pathology enables us to add a very important qualification—namely, that most of the cases in which formidable cerebral symptoms have been removed by the action of strong purgatives, have been neither inflammatory nor apoplectic in

their nature, but cases in which brain-symptoms have resulted from blood-poisoning; and in the majority of instances the poison has been uræmic. Amongst the cases
recorded by Dr. Abercrombie as examples of inflammation
of the brain successfully treated, there are two (Cases LXX
and LXXX) in which the brain-symptoms were associated
with anasarca, which had followed an attack of scarlet
fever. These were unquestionably cases of acute renal
disease, with cerebral symptoms of uræmic origin. I offer
for your practical guidance this rule of treatment : when
such symptoms as headache, delirium, convulsions, or coma
are the result of uræmia, give purgatives freely ; and, if
the renal disease be acute, and therefore probably curable,
your treatment will often be completely successful. On
the other hand, when you have reason to believe that
the like brain-symptoms are consequent on cerebral
hæmorrhage, or embolism, or thrombosis, be very cautious
in the use of purgatives, which may greatly increase the
patient's distress and exhaustion, while they can do little
to improve his condition. In inflammatory affections of
the brain and its membranes, purgatives are often useful,
but less frequently and strikingly so that when cerebral
symptoms are the result of uræmia. As to the form of
purgative in uræmic cases, croton oil is the most convenient when there is coma and consequent difficulty in
swallowing a more bulky dose. When there is no such
difficulty, two pills, composed of three grains of calomel
with seven grains of compound colocynth pill, may be followed in four hours by an ounce of the compound senna
mixture; or the following powder, which I think an improvement on the compound jalap powder, may be given ;
℞ Scammoniæ resinæ gr. v. to viii ; potassæ tartratis
acidæ Əi ; pulveris zingiberis gr. iii. Misce. The dose
to be repeated once or oftener, according to circumstances.

When, in a case of acute Bright's disease, the renal
congestion is excessive, as shown by the scanty secretion of

highly albuminous urine, with vomiting, headache, and other threatening nervous symptoms, local bleeding by leeches or cupping on the loins is often extremely useful, and is quickly followed by an increased secretion of urine. If, by the abstraction of a few ounces of blood from the loins, we relieve the renal congestion, we thereby lessen the destruction of blood-constituents which results from contamination of the blood by urinary excreta. Moderate and timely local bleeding, therefore, tends to economise blood, and prevent its waste.

It has been asserted that cupping or leeching the loins can help an inflamed kidney no more ' than if the blood had been taken from the arm or from the nape of the neck. But this, surely, is a mistake. The lumbar arteries, which supply the integuments of the loins, arise from the abdominal aorta, close by the origin of the renal arteries; and, when leeches or cupping-glasses draw blood through the skin of the back, it is certain that the diminished pressure within the lumbar arteries will divert a certain quantity of blood from the neighbouring renal arteries. The same principle explains the good effects of leeching in cases of pericarditis. The internal mammary artery sends deep branches to the pericardium, and superficial branches to the intercostal spaces and the skin. By the application of leeches over the heart, we abstract blood from the integumentary branches of the internal mammary artery, and thus divert a portion of blood from the deeper pericardial branches. The blood will as surely take the course indicated by diminished pressure within the vessels, as the water in a pump will, up to a certain height, follow the rising piston. It may be thought that the quantity of blood thus diverted is very small; so when venesection is practised in the arm or neck, how scanty is the stream of blood which escapes from the opening in the vein, compared with the torrents of blood rushing through the venæ cavæ into the right side of the heart; and yet, in a case of obstructed circulation through the heart or lungs, how promptly and decidedly does this small diverted current lessen the distension of the whole venous system. As a rule, I prescribe local bleeding only when, the secretion of urine being extremely scanty, there is a consequent threatening of head-symptoms or other serious rusults of uræmia. In ordinary cases, I apply hot fomentations or poultices covered by mackintosh to the loins. These act by relaxing the superficial arteries. The skin, therefore, receives a larger supply of blood, and thus a portion of blood is diverted from the renal arteries. Then, too, there is some degree of depletion from the full cutaneous capillaries by the free local sweating which the warmth occasions.

Dry cupping acts in a somewhat similar way to hot fomentation. It draws an abundance of blood through the arteries into the subcutaneous capillaries, which, when the cups are removed, returns through the veins to the heart. In order that dry cupping may be most effectual, each cup should be removed as soon as the vessels beneath are well filled, and then it should be reapplied. The object is first to draw the blood through the arteries into the capillaries; then to allow it quickly to return by the veins, and not to keep it stagnating in the capillaries, which will happen if the glass be retained long on one spot. Another point is not to draw the blood into the skin with sufficient force to cause extravasation, the effect of which will be to impede the circulation through the skin, and so to divert more blood into the inflamed tissues beneath. The sole object of dry cupping, be it remembered, is not to irritate the skin, but to draw blood rapidly from the arteries, and as rapidly to transmit it through the capillaries to the veins, in its backward course to the heart.

As a rule, it is well to give no alcoholic stimulants; or, if need be, to give them very sparingly in cases of acute Bright's disease. The imbibition of alcohol imposes extra work on the kidney, and so is opposed to the principle of lessening as much as possible the work of the inflamed gland. Excess

of alcohol is, amongst the lower classes, one of the most
frequent causes of albuminuria ; and a very moderate employment of alcohol may tend to perpetuate and aggravate
disease originating from other causes.

When acute Bright's disease is making satisfactory progress
towards recovery, the dropsy usually disappears for a variable
time before the urine ceases to be albuminous. It is very
important to impress upon the patient that, until 'the urine
has regained its normal characters, he should be warmly
clothed with woollen next the skin ; and he must be extremely careful to avoid cold, fatigue, and errors of diet.

The duration of albuminuria in cases that ultimately
recover is very variable. I have seen many cases of recovery
after the disease had continued from three to twelve
months, and I have seen some recoveries after the urine
had been albuminous for one, two, and even three years.

The more I have seen of the disease, the more hopeful
I have become as to the ultimate result, when the history
and the symptoms, and, above all, the chemical and microscopical characters of the urine, do not indicate extensive
and irremediable degeneration of the kidney. In all the cases
of recovery from long-continued albuminuria, the preparations of iron have entered largely into the medicinal treatment of the disease, and have apparently contributed much
to the favourable result. There are two preparations which I
have found especially useful ; these are the tincture of the
perchloride and the syrup of the phosphate—the former in
doses of from ten minims to half a drachm, and the latter
in drachm doses twice or thrice daily. The preparations
of iron are best taken soon after food. I have frequently
combined with each dose of the perchloride of iron ten
grains of hydrochlorate of ammonia ; and I believe that
this ammonio-chloride of iron is an useful combination. The
preparations of iron should not generally be given after the
use of tea as a beverage, for the reason that the resulting
tannate of iron is apt to irritate the stomach, and is less

readily absorbed. In most cases it is well to omit the iron every third day, the patient being thus enabled to continue the medicine longer, and to utilise it more completely. If the full doses of iron which I have recommended are not well borne, smaller doses should be cautiously tried. It is especially important to bear in mind that any medicine which causes disorder of the digestive organs will be injurious rather than beneficial.

Amongst other remedial agencies, when acute renal disease is prolonged and threatens to become chronic, change of air and scene is often highly beneficial. Residence during the winter season in a warm, dry, equable climate, such as may be found at Cannes, Nice, and Mentone, has in many instances been attended with highly beneficial results. The bright warm sun, and dry invigorating air favour the action of the skin and of the bronchial mucous membrane, the patient is able to be much in the open air, and thus the respiratory, the digestive, and the secretory functions are all assisted and promoted. I have seen some most remarkable recoveries effected under the influence of a long voyage after other means had failed to effect a cure.

The treatment of *chronic* Bright's disease must obviously vary according to the form and stage of the malady and the nature of the secondary complications. In each case, it is practically important to ascertain, if possible, the probable cause of the renal disease. Your inquiries should be directed to determine whether the chronic malady is a sequel of an acute attack, or whether it commenced as an insidious chronic disease. Then you should make inquiry as to the exciting cause, which in most cases may be arrived at with a high degree of probability. Is there a history of gout, or of habits likely to induce a gouty diathesis? Excessive eating and drinking, chronic dyspepsia, frequent exposure to cold and wet, cachexia the result of syphilis or of other constitutional disease, scrofula or other possibly hereditary taint, and chronic lead-poisoning, are amongst the probable determining causes which should be made the subject of inquiry; and then the treatment should be directed to remove, if possible, or, so far as may be, to counteract, the morbific influence.

Whatever may be the form or the stage of chronic Bright's disease, the skin must be protected from cold by warm woollen clothing, care must be taken to avoid over-fatigue, and the diet should be carefully regulated both as to quantity and quality. As a rule, in all cases of chronic renal disease, alcoholic stimulants in any form should be given sparingly, or abstained from entirely, unless, for some special reason, they appear to be indicated. You will find that, when there is extensive degeneration of the kidney, alcoholic liquors usually produce far more decided, and often deleterious effects, than result from equal quantities of the same liquors when the kidneys have their normal structure and functional activity.

In one class of cases—cases of large white kidney, with a scanty secretion of highly albuminous urine—*dropsy* is usually a prominent symptom, and requires special treatment. The tendency to dropsy is no doubt increased by the dry and inactive state of the skin, which often resists the relaxing effect of external warmth, so that a hot-air bath, or even the hottest room of a Turkish bath, fails to excite diaphoresis. Patients who do not perspire under the influence of the hot-air bath, usually complain of painful throbbing in the head, difficult breathing, and other distressing symptoms. On this account, I prefer in most cases the wet sheet and blanket bath, which, as a diaphoretic, is both more agreeable and more efficacious than the hot-air bath.

Diuretics are notoriously uncertain in their operation. In order to assist the action of diuretics, diluents should be freely given; and I have often obtained most satisfactory results by keeping the patient entirely on milk, either skimmed or unskimmed, in accordance with the rules which I have before laid down, with the addition of a mixture containing acetate of potash and infusion of digitalis, a drachm of each for a dose, to be repeated three times a day. By these means, a copious secretion of urine is induced, and the dropsy is speedily and completely removed. A strong infusion of fresh broom-tops, taken in sufficient quantity every morning to act as a purgative, often proves a very efficient diuretic. The *succus scoparii* of the *Pharmacopœia* may be substituted for the fresh infusion. The imperial drink (cream of tartar and lemon), in doses of from two to four pints in the twenty-four hours, is also a pleasant and efficacious diuretic. The late Sir James Simpson was the first to use the vapour of oil of juniper as a diuretic. Thirty or forty drops of the oil may be floated on boiling water in an ordinary jug or in a suitable inhaler; and the mixed vapours of the volatile oil and water may be thus inhaled twice a day. In some cases, the diuretic action is very prompt and decided. Another plan, originally proposed by Dr. Christison, is to apply digitalis freely to the skin. Dr. Christison's plan consists in making a strong infusion by adding an ounce of the dried digitalis-leaves to a pint of boiling water. A large piece of spongio-piline, steeped in the infusion, is kept constantly applied to the abdomen. An alternative plan is to pour an ounce of tincture of digitalis on the surface of a large hot linseed-poultice, which is then applied over the loins and back; the poultice, with the tincture of digitalis, being renewed two or three times a day. This plan sometimes succeeds when other methods have failed to remove the dropsy.

The free action of a hydragogue purgative, such as elaterium, compound gamboge pill, compound jalap powder, or the powder composed of scammony resin, cream of tartar, and ginger, which I have before mentioned, is often followed by a more copious secretion of urine. The probable explanation of the indirect diuretic action of a hydragogue purgative is this. The purgative excites a copious watery secretion from the blood into the bowel; this is followed by the absorption of a portion of the dropsical fluid which had been effused into the areolar tissue, and perhaps into one or more serous cavities. The partial absorption of the dropsical effusion removes or lessens the pressure on the vessels, more especially on the veins; and so the circulation becomes more free, at the same time that the absorbed liquid exerts a diuretic influence on the kidneys similar to that which we have seen to occur during convalescence from acute renal dropsy, and not unlike that which often results from the introduction of abundant diluents through the stomach.

When other means fail to remove the dropsy, when the anasarcous distension of the legs is increasing and causing pain and incipient erythematous inflammation, or when the breathing is becoming impeded by the accumulation of water within the abdomen or the chest, or by an œdematous condition of the lungs, prompt, decided, and sometimes permanent relief may be afforded by allowing the water to escape through an incision in the skin, about half an inch long, just above either the outer or the inner ankle of each leg. The incision must be deep enough to enter the areolar tissue beneath the skin. You are aware that, for incising dropsical legs, I am in the habit of using a small instrument made for me by Messrs. Weiss. It may be described as a cupping scarifier with a single blade, which, on touching a small knob, is thrust out by a strong spring, and thus makes a clean cut through the skin with such rapidity that it causes little or no pain. We had lately a good opportunity of testing the comparative painlessness of an incision made by this instrument. A dropsical patient had his legs acupunctured by the house-physician, and cried out with the pain caused by the needle-punctures. A few days afterwards, the punctures having ceased to discharge, while the dropsical swelling was but little reduced, we made an incision into each leg with the spring lancet. He declared that he scarcely felt the cuts ; and the incisions discharged so freely, that the dropsy was for a time completely removed. I have seen many cases in which life has been prolonged for a considerable period, and some in which a complete and permanent cure has followed incision of the legs, after other means had failed to afford relief. To refer to one case only out of a number : towards the end of July 1861, I first saw a clerk to the New River Company, aged 22. Since the end of March, he had suffered from general dropsy, the result of exposure to cold. The urine became nearly solid with heat and acid, and it contained numerous oily casts. Purgatives and diuretics failed to lessen the dropsy ; and at the beginning of September the swelling of the legs was so great that the skin cracked, and water oozed through the fissures. I then advised that the legs should be incised. A copious discharge of water occurred, and the urine became more copious. From that time he steadily improved, the dropsy passed away, and gradually the urine ceased to be albuminous ; but it was not until the end of April 1862, more than a year from the commencement of his illness, that all trace of albumen had disappeared. The chief medicinal treatment after the incision of the legs consisted in giving the tincture of perchloride of iron three times a day, and a dose of strong broom-tea every morning. Since his recovery he has insured his life, he has married, and has several children. I heard of him quite recently as remaining well. I have thought this case worthy of especial mention, as an example of complete recovery after dropsy to an extreme degree, and albuminuria with numerous oily casts, had continued for the greater part of a year; the first favourable change in the patient's condition following directly upon a copious discharge of water through incisions in the legs.

After the dropsical legs have been punctured, the folded sheet and mackintosh, placed beneath to receive the serous discharge, should be frequently renewed, and kept clean. The liquid quickly decomposes and becomes ammoniacal, and in this state it may irritate and inflame the skin. Cleanliness is, therefore, essential for safety as well as for comfort. Any inflammatory redness about the wound may usually be removed quickly by the application of a lead lotion. It is true that severe inflammation and sloughing have sometimes followed incisions or punctures in anasarcous legs; but this may, and often does, occur from over-distension of the skin, or from the mere pressure of the heavy dropsical limbs upon the bed. The result of my experience is that inflammation of anasarcous legs has been as often subdued as provoked by acupuncture or incision; that inflammation is much less likely to follow incisions in cases of renal than of cardiac dropsy, when the circulation is much impeded by valvular disease; and that an incision made with the spring scarifier is as safe as acupuncture, and much less painful.

The copious secretion of urine which usually follows a discharge of dropsical fluid through incisions in the skin admits of nearly the same explanation as that which I have already given you of the like phenomenon after the action of a hydragogue purgative. The escape of the dropsical effusion through one or more incisions in the skin, removes pressure from the veins, and permits the blood to move more freely through the vessels. This greater freedom of the circulation is attended by a quickened absorption, and some of the absorbed dropsical liquid, more or less charged with urea, enters the circulation, and exerts a diuretic influence on the kidney. In the treatment of a copious dropsical effusion the main object, and the chief difficulty, is to overcome the *vis inertiæ* of the stagnant liquid. If once we can set the liquid in motion, whether by a primary diuretic action upon the kidneys, by first exciting a free discharge of liquid through the bowels, or by giving exit to a portion of the liquid through incisions

in the skin—in whichever way the current is started—
the outward movement of liquid often continues until the
whole of the dropsical effusion has been swept away, and
in each case a free secretion of urine constitutes a part of
the eliminative process.

The *anæmia* of chronic Bright's disease is to be counteracted by a carefully regulated diet, and by the persevering
use of one or other of the preparations of iron. When with
anæmia there is a scanty secretion of urine, and a tendency
to dropsy, a very useful combination is twenty minims of
the tincture of perchloride of iron, with a drachm of spirit
of nitrous æther, and a drachm of infusion of digitalis, or
from ten to twenty minims of the tincture of digitalis in
an ounce of water, given three times a day, soon after food.
If the mixture be found to constipate, from half a drachm
to a drachm of sulphate of magnesia may be added to each
dose, or the bowels may be acted upon by an occasional
dose of the compound colocynth pill. Mercury in any form
often acts powerfully and injuriously in cases of Bright's
disease. The chief use of mercurials in these cases is to
assist the operation of saline or vegetable purgatives.

Syphilitic symptoms, when present, are best treated by
gradually increasing doses of iodide of potassium, with
bark, or quinine. The calomel vapour bath has been found
a useful remedy in some syphilitic cases.

When chronic renal disease is associated with a *strumous
diathesis*, cod-liver oil may often be given with advantage.

Dyspœna is one of the most frequent and distressing
symptoms associated with advanced Bright's disease. It
has various causes, and requires various remedies. When it
results from œdema of the lungs, or dropsical hydrothorax,
it is best treated by the remedies for dropsy. In some cases,
anæmia appears to be the chief cause of dyspnœa. The red
blood-corpuscles are the oxygen-carriers. When the blood—
whether in cases of chlorosis or of Bright's disease—contains
an excess of water with a corresponding deficiency of red

corpuscles, the defective oxidation of the tissues and the demand for air are manifested by hurried and laborious breathing. The remedy for this form of dyspnœa is to be sought for in the elimination of water, a carefully regulated nutritious diet, and iron as a restorative tonic.

When dyspnœa results from pulmonary congestion and bronchial catarrh, it is best treated by warm baths, fomentations, or poultices to the chest, and mild counter-irritants.

Paroxysmal dyspnœa, in many cases of advanced Bright's disease, appears to result from the influence of deteriorated and poisoned blood upon the nerves and the nervous centres. The heart's action is rapid and feeble, the breathing is distressed and hurried, while the respiratory murmur is loud, puerile, and unattended by wheezing or crepitating sounds. This form of dyspnœa is usually more common and more distressing at night. It is not improbable that, in some cases, cardiac weakness, with palpitation and dyspnœa, may result from excessive contraction of the minute branches of the coronary arteries, excited by the stimulant action of morbid blood, and causing anæmia and malnutrition of the muscular walls of the heart. These symptoms are most effectually warded off by a carefully regulated diet, by promoting the action of the skin and bowels, and so far as possible of the kidneys, the object being to free the blood from accumulated impurities. Temporary relief is often afforded by ether, or by brandy. In some cases, a small dose of chloral hydrate, not more than ten grains, repeated at intervals of six or eight hours, is very beneficial. The chloral has this advantage over every preparation of opium, that it has no astringent action on the bowels, and that it rather increases than checks the secretion of the kidneys. It therefore never excites the distressing sickness which often results from the astringent influence of opiates in the advanced stages of renal degeneration. The *muscular twitchings*, and the *painful cramps*, which are common results of uræmia, and not unfrequent precursors of convulsions, may sometimes be kept off, or much mitigated, by twenty-grain doses of bromide of potassium, given twice or three times in the twenty-four hours. A combination of chloral hydrate with bromide of potassium appears sometimes to have a powerful influence in warding off uræmic convulsions.

Although the use of *opium* in all forms and stages of Bright's disease requires extreme care, on account of its tendency to check all the secretions except that of the skin, yet you will occasionally meet with cases in which the distressing nervous symptoms resulting from uræmia are relieved by opiates more effectually than by any other means. There are cases in which the carefully-observed result of a cautiously conducted experiment is sometimes a better guide than theory.

The sufferers from Bright's disease are always *dyspeptics*, and the *gastric symptoms* are often very obstinate and distressing. When, in consequence of renal degeneration, the blood is contaminated by retained urinary excreta, there is often a vicarious excretion of these impurities by the mucous membrane of the stomach and bowels. The gastric secretions are mingled with the ammoniacal products of decomposing urea; digestion is consequently impaired; there are flatulent distension of the stomach and bowels, nausea, vomiting, and diarrhœa. Relief is to be sought by a carefully regulated diet, and by giving with the food from ten to twenty drops of dilute hydrochloric acid with a vegetable bitter. The liquor strychniæ in doses of five minims, or the tincture of nux vomica in ten-minim doses, with a mineral acid, is sometimes especially efficacious. Pepsine may sometimes be given with advantage.

In some cases of advanced renal degeneration, the *vomiting* is so incessant, that the patient has to be sustained by nutritive enemata, while iced water only, or iced milk in small quantity, is taken by the stomach. In some instances that have come under my observation, the straining and exhausting efforts of vomiting have been checked only by frequent whiffs of chloroform vapour.

When the stomach is very irritable, chloral can rarely be retained, it acts at once as an emetic; but injected into the rectum, its soothing influence is very similar to that of chloroform vapour, while it has the advantage of producing a more durable impression on the nervous system, and therefore requires to be less frequently repeated.

When the retina is the seat of hæmorrhage, or of albuminuric retinitis, the eyes should be allowed to rest, and they should be carefully shaded from excess of light. Recovery of sight, more or less complete, may occur, but it is doubtful whether treatment has much influence upon the symptoms. My colleague Mr. Soelberg Wells, with whom I have seen several cases of this affection, strongly recommends moderate local bleeding by the artificial leech. When, on account of the anæmic condition of the patient, the abstraction of blood is undesirable, he has seen marked improvement follow the application of the dry cup to the temple, at intervals of five or six days. ('A Treatise on the Diseases of the Eye.' By J. Soelberg Wells. Second edition, p. 361.)

INDEX.

www.ingramcontent.com/pod-product-compliance
Lightning Source LLC
Chambersburg PA
CBHW021809190326
41518CB00007B/523